FOLKLORE OF THE OIL INDUSTRY

by
Mody C. Boatright

with illustrations by
William D. Wittliff

SOUTHERN METHODIST UNIVERSITY PRESS • DALLAS

Library of Congress Cataloging in Publication Data

Boatright, Mody Coggin, 1896-1970.
    Folklore of the oil industry. With illus. by William D. Wittliff.
    Dallas, Southern Methodist University Press [ᶜ1963]
    vii, 220 p.   illus.   24 cm.
Bibliographical references included in "Notes" (p. 205-213)
    1. Petroleum—U. S.—Hist. ɪ. Title.
TN872.A5B6   622.3382   63-21186 rev
ISBN 0-87074-007-5
ISBN 0-87074-204-3 (paper)

Preface

THE OIL INDUSTRY is a little more than a hundred years old—old enough to have generated a considerable body of tradition, young enough to exemplify the generation of tradition in a literate, industrial society. It is the purpose of this book to examine some of the more prevalent forms of this tradition—to describe and illustrate some of the basic patterns under which the folk have structured the experience relating to the seeking for and the production of petroleum.

By the folk I mean simply the creators and carriers of tradition —those people of whatever level of sophistication whose fanciful reaction to the conditions and events about them results in creation, and those who have sufficient interest in what they have seen and heard (and in some instances read) to repeat it and thus place it in the oral tradition.

The carriers of tradition are also creative in that memory is selective and that conscious or unconscious reshaping of data is inevitable. For it is a commonplace that events do not express their own significance; that the mind rebels against disorder and chaos; and that all learning and all art are efforts to impose upon the data of experience a structure that will give them balance and order—that is, make them intellectually or emotionally satisfying. In this respect folklore resembles both art and learning. When confronted with novelty, the folk, like the artist and the scholar, attempt to relate it to the familiar. Thus new traditions are engrafted upon old ones, the old revised to accommodate the new.

My interest in oil lore began many years ago when I first discerned certain patterns common to various oil regions in Texas.

I began filing notes, and later was able to make a systematic investigation. My field work has been done principally in Texas, but it soon became apparent that much lore had migrated from the older oil states. I accordingly attempted to familiarize myself with the oil lore of these states through research in libraries and archives and through field work in Pennsylvania, West Virginia, Ohio, Kansas, and Oklahoma.

For the terminal date of my study I have chosen 1940 as marking, as accurately as such things can be marked, the end of the old techniques and the passing of the industry to a new generation.

Since the Texas oil millionaire had not by that date emerged clearly as a nationally known folk character he is omitted from the list of stereotypes, though I have inconsistently included a few anecdotes about him.

I wish to express my indebtedness to the following:

To the University of Texas for grants in aid that made this book possible.

To Mrs. Walter Sharp and her son Dudley Sharp, who financed the University of Texas Archive project known as the Oral History of Texas Oil Pioneers, in which I was privileged to participate.

To Miss Winnie Allen, former archivist of the University of Texas, who in many ways facilitated this study.

To nearly two hundred informants who permitted their reminiscences to be tape recorded and filed in the Archives.

To many other friends and informants who have helped in many ways.

To the *Southwest Review, Lamp*, the *Texas Quarterly*, and the Texas Folklore Society for permission to reprint portions of this work previously published in the journals named and in the Publications of the Society.

MODY C. BOATRIGHT

Austin, Texas
August 12, 1963

Contents

I

Finding Oil

THE BLENDING *of the old and the new is especially apparent in the folklore relating to the search for oil. Men have for ages looked for techniques for finding water and minerals under the ground, and the lack of pragmatic evidence of the efficacy of these techniques did not impede their use, nor prevent the engrafting of pseudoscience upon them. Indeed, since fortunes were made by the discovery of oil, and since science could not with certainty tell a man where to drill, the pressure to resort to superstition and pseudoscience was compulsive; and those who made specious and not always plausible claims to infallibility did a profitable business, and sometimes by accident served their clients well.*

1

There's Oil
Under Them Hills

STANLEY VESTAL once asked a petroleum geologist why he did
not expose the doodlebug men and other fakirs who were com-
peting with him for the fees of the oil producers. "Not me,"
replied the geologist. "It would ruin my profession. More oil has
been found by doodlebugs than regular geologists."[1]

Since this statement was made, geological knowledge has
expanded, and the seismograph, the torsion balance, and the gravity
meter have come into wide and successful use in mapping hidden
strata. Independent wildcatting has declined and with it the pick-
ings of the oil witch. The geologist Vestal quoted in 1941 might
not wish to reaffirm his statement today, but he would not mini-
mize the importance of folk-belief in the history of the oil industry.

Drake's well, completed in 1859, the first in the United States
drilled specifically for oil, was located between the margin of a
creek and a steep acclivity, from which facts the deduction was
made that oil was to be found only between a creek and a bluff
or steep hill, and large areas that later proved productive were
bypassed because of the absence of streams and bluffs.

But dreamers and seers and diviners, men of imagination as
well as of vision, refusing to be bound by the restrictions of creekology, as the rudimentary science of petroleum geology was then
called, made locations wherever their intuition or their instruments
led them; they made locations on other kinds of terrain, and
on some of these locations drilling brought oil and with it new
facts for science to digest. As late as 1942 no less distinguished

3

a geologist than Wallace Pratt noted that geologists like to emphasize their contribution to the oil industry, but that we "might agree that our effort to make geology serve industry has benefited geology more than industry."[2]

Thus in many respects the oil witch bears the same relation to petroleum geology and geophysics that the alchemist bears to chemistry.

Both petroleum geology and folklore begin with a search for surface indications of oil. One such indication utilized by oil-seekers was plant life. Some practitioners merely looked for the species found in already proven fields, but this was too easy for the pretenders to esoteric learning. They sought specific plants, which would of course vary with the region. It has been said that M. A. Davy, the discoverer of the Boggy Creek field, was attracted to the dome "by a saline prairie which grew a grass that cattle refused to eat."[3] The grass suggested subsurface salt, which suggested a salt dome, one of the common oil-bearing formations on the Texas coastal plain.

For the charlatan the problem was not an easy one. It was difficult for him to observe every foot of ground and to note every plant; and the plant chosen must be a relatively rare one, yet one that did exist in the region. One oil-seeker in the Ranger field based his prognostications upon the shoestring cactus. The significant thing, however, was not the mere presence of this cactus but the direction in which the leaves pointed. If they pointed east, you could safely drill.

I have heard only one man try to rationalize the technique of finding oil by the presence of plants, and he had a scientific explanation. Over an oil pool, his theory ran, evaporation, even through the most dense strata, would give the topsoil more than the normal content of hydrocarbon. You would then look for plants that demanded a heavy hydrocarbon diet. Naturally, to tell what these plants were would be to betray a trade secret and ruin his own business.

Some sanction for his theory might be seen in the work of certain geochemists who sought oil, some with considerable success (but whether due to luck or science is a moot point), by analyzing the topsoil, a high content of hydrocarbons being an indication of the presence of oil.*

Any unusual phenomenon, especially in retrospect, might be associated with the presence of oil. One man, after the discovery of oil in Anderson County, connected the pressure of oil and gas with the sagging of a door. He said, "I have noticed that in my residence the door sags one way a part of the year and another way the balance of the year." He said that various theories were advanced to explain this phenomenon, one being that it was caused by a subterranean pool of oil.[4] In Desdemona, another man had long suspected the presence of oil because sound traveled so far. He could hear the train in Thurber twenty-five miles away. It wasn't reasonable for sound to travel like this unless there was some underground passage through which it passed.[5]

The oil industry was still young when geologists, abandoning creekology, began to favor anticlines as locations for drilling. To the layman this meant that oil was to be found in high places— hills and mounds of all descriptions—and he assumed that the high places had been lifted by gas pressure. Thus farmers and ranchmen in Central and West Texas became convinced that vast pools of oil underlay their land, although as a matter of fact most of their hills were remains of eroded plateaus. They found corroborating evidence in the presence of numerous small mounds containing burnt stone of a pinkish color. These, they said, were places where in past ages gas had broken through to the surface and become ignited. Archeologists refer to these mounds as kitchen middens.

One of the most widespread and persistent folk beliefs is that oil is likely to be found in or near a graveyard. This belief is sup-

*For a description of one such process see the *New York Times* for March 16, 1941.

ported by considerable empirical evidence, for there are few popu-
lated oil regions where oil has not been found in or near a graveyard. Some geologists believe that more is involved than mere
chance. High ground is usually selected for the cemetery, and
high ground sometimes means an anticline or a salt dome, and
anticlines and salt domes sometimes mean oil.

Drilling in cemeteries presented some delicate moral and legal
issues. For hundreds of years custom and common law had pro-
tected burial places from desecration, but neither custom nor law
was explicit on the subject of drilling oil wells. Two cases from
the Ranger field illustrate the confusion that arose.

In that field practically all the privately owned land had been
leased by one company before the drilling of the discovery well
on McClesky's farm. For latecomers there was little left except
schoolyards, churchyards, and cemeteries. One operator put down
a well on the Merriman school playground that made the district
in per capita wealth the richest in the world. The twenty-nine
members of the Merriman Baptist Church then voted to lease the
churchyard. The well was a gusher. By unanimous vote the congregation set aside 15 per cent of their great wealth for the support
of their church and gave the remainder to such causes as home
and foreign missions, hospitals, orphanages, and colleges. Not one
member profited individually.[6]

Operators next sought the right to drill in the Merriman
Cemetery, but although they offered a cash bonus of $100,000,
the trustees refused to sign the lease. For this they were widely
applauded, and Will Ferrell was so impressed by their stand that
he wrote a poem which has become a part of the folklore of
the oil fields:

> All oildom knows the answer,
>   When the chairman shook his head,
> Pointing past the men of millions
>   At the city of the dead....

> "Why disturb the weary tenants
>   In yon narrow strip of sod?
> 'Tis not ours but theirs the title,
>   Vested by the will of God."

> "We the board have talked it over,
>   Pro and con without avail,
> We reject your hundred thousand—
>   Merriman is not for sale."[7]

But the picture of the deacons heroically "standing guard" above the gravestones in a lot that's "not for sale" was somewhat toned down by the statement of the minister to C. C. Rister, that the church could not have made a valid lease "even if it had desired, for the donor [of the cemetery] had deeded it to the dead."[8]

The trustees of the Pleasant Grove Cemetery, also near Ranger, upon authorization of the Methodist Conference called specifically to act on the matter, exchanged an unused part of the cemetery for stock in the Hazel Fain Oil Company, which was to put down a well on the tract. R. E. Barker, feeling that this act would constitute desecration of his father's grave, went to court seeking to enjoin the drilling. His petition alleged that the defendants were "now threatening to erect derricks, dig slush pits, etc., for the drilling of an oil well, which . . . would inevitably result, if allowed to proceed, in discordant noises, obnoxious odors and the desecration of the graves by the spraying of oil."

He secured a temporary injunction, which was dissolved by the trial court, and the derrick went up. One corner of it was within inches of the elder Barker's grave. The son then appealed to the Court of Civil Appeals, whose decision, affirmed by the Supreme Court, became a precedent for other cases and clarified the law in Texas. The right of relatives of dead persons to sue was established. The injunction was made permanent. The decision did not prohibit the drilling of oil wells in cemeteries, but it made it necessary for one seeking the right to drill to secure the permission of all the descendants of all the deceased and their near relatives. This of course is a practical impossibility in the case of old or large cemeteries. The law seems to say that the act itself is not desecration unless some interested person thinks it so.[9]

So it was possible for a reporter to send in the following idyllic picture from Gladewater in 1934:

> Surrounded by tombstones a spouting oil well flowed today in the center of Roosevelt Cemetery, one mile east of here.
>
> Brought in by the Circle Oil Company of Houston, the well, under heavy gas pressure, flowed beside numerous graves overshadowed by tall pine trees.
>
> Cemetery officials said their burial ground was probably the only one in the United States that is self-supporting, with a treasury fund.[10]

By 1937, when a cemetery case reached the court of Judge Minor Moore of California, the technique of slant drilling had been perfected so that it was possible to drain the oil from under the dead without setting up derricks and driving trucks over their graves. Mrs. Kate McCullough and ten other plaintiffs described as descendants of deceased persons buried in Sunnyside Cemetery filed suit for damages in the amount of $226,000,000 in behalf of themselves and 2,500 other persons. Judge Moore held that the facts proved no disrespect to the dead and afforded no grounds for damage. "The cemetery," he declared, "is all in order, row on row; not a leaf has been stirred or a flower uprooted. Serenity and calm prevail within the grounds."[11]

The protests all came from the living. There is no tradition that the dead in any cemetery where drilling was carried on spoke up in rebuke of those who disturbed their rest. But a driller, known to my informant only as Hardboiled—a name suggesting the probable attitude of his helpers toward him—might, were he asked, give conflicting testimony. He was one of the men who drilled in Pleasant Grove Cemetery. He set the first string of pipe and then knocked off for four or five days to wait for the cement to harden. While he was away some of the workers who were living in a tent nearby rigged up a string of two-inch pipe, leading underground from their tent to the cellar of the well. When Hardboiled returned and started drilling, they began hollering through the pipe, "Oh Lord! My God! Have mercy. Have mercy on us." Hardboiled became more and more nervous, and after about an hour of groans and pleas for mercy he said he was not going to drill another lick. He walked off the job and another driller brought in the well.[12]

The men who tried to secure the right to drill on church lands and burial grounds were hardheaded operators with no more piety than other men. They were concerned about titles but not about owners. There were pious men, however, who seemed to think that the Lord would be disposed to reward a search for oil on holy ground. When a well was to be spudded in the yard of a country church near Longview in 1931, the pastor called his flock together for a "prayer meeting to ask divine favor on the project." A month later "when the liquid wealth came swelling up, he hastily summoned his people for a service of thanksgiving and praise."[13]

There were some promoters with interests or stocks to sell who said nothing to discredit the pious belief that there was some relation between the wealth under the soil and the worthiness of the titleholder. This idea is prominent in the promotional literature of the notorious Frederick A. Cook, whose oil activities finally landed him in a federal prison.

But poor widows figure more prominently in such promotional activities than churches. One trick was to plant, previous to the stock-selling campaign, news stories suggesting the intervention of Providence on behalf of the lowly. I can think of no other hypothesis to account for the following item in the *Houston Chronicle* for January 26, 1910:

DALHART, TEXAS, Jan. 25—Mrs. Josie Pettie, owning a farm near here, today declared a bolt of lightning struck the mountain side near her home and uncovered a spring of crude oil now producing 200 barrels daily. She has been offered $10,000 for a small tract that had been barely yielding her a living.

A columnist in the *Oil Investor's Journal* commented:

Since an Oklahoma widow appeared in Guthrie and took out a charter for an oil company, declaring that lightning had uncovered petroleum on her property and that she would drill and develop the land, there has been quite an epidemic of such pranks by the elusive electric force, if the daily papers are to be believed.... It must make some of our wildcatting friends, like Walter Sharp, awful sore to see the lightning going around in a blind sort of way turning up oil pools here and there, while the boys with the drill are not able to get even so much as a smell. We suggest that Walter harness the lightning and make it work for him. Either that or make some sort of deal with the widows.[14]

He might have added that there are no mountains and therefore no mountainsides for the lightning to strike near Dalhart.

If you were looking for oil, then, you should not neglect hills and mounds; you should if possible secure leases on graveyards and church property and poor farms of poor widows. But when you had attended to these details, you had by no means exhausted the techniques of finding oil. Hosts of seekers, amateurs and professionals, were ready to assist you with methods frankly occult or reputedly scientific as you might prefer.

2

Dreams of Oil

APPARENTLY oil has been found as a result of dreams, and dreams have presaged the discovery of oil without being the immediate cause of drilling. These things happen, however, I am convinced, less frequently than reported. It has often been said that this man or that drilled on a certain location because he had dreamed that oil would be found there. The man is not available, and nobody knows the content of his dream. Persons who knew him cannot recall that he ever mentioned a dream in their presence. Perhaps there has been transference: one person's experience has been transferred to another. Perhaps a metaphor has been given literal interpretation.

It is widely believed and has been stated in print that Mrs. S. L. Fowler's refusal to consent to the sale of the Fowler farm near Burkburnett until a test well had been drilled was due to a dream she had repeatedly dreamed. Yet no informant has been able to reveal exactly what her dream was, what events she experienced or what images she saw in her dream.[1]

There is no accepted symbolism about dreams significant of petroleum. One still hears occasionally the old belief that dreams go by contraries—that to dream of a funeral is a sign of a wedding, and that to dream of gold is a sign of poverty. By this reasoning a producer ought to be happy if he dreams of a dry hole, but I have never heard of anybody's drilling for oil on a location because he had dreamed oil wasn't there. There is also a belief that to dream of oil means success, satisfaction, good crops. The saying is older than the petroleum industry and probably refers to olive oil. Some dreams that have been cited as indicating the presence of oil under the ground have been literal and direct. That was the kind a traveling salesman in northeastern Indiana had some years ago. He dreamed of a producing well between two dry holes that had supposedly condemned the territory. He had faith in his dream. But when he told it to the oilmen, they laughed at him. No producer would risk his capital on a dream. The salesman formed a partnership with some other amateurs and opened a pool. Then he laughed at the oilmen.[2]

Years earlier, in 1864, a clerk by the name of Kepler in Hydetown, Pennsylvania, had a dream. In his dream he was wandering through the woods with a young lady, who was, according to a contemporary, "considered somewhat of a coquette." As they wandered pleasantly through the woods, Kepler suddenly noticed some distance away an Indian with bow bent ready to shoot him. The young lady handed him a rifle, which she had not had a moment before; he took it and fired at the Indian, who immediately vanished. But on the spot where he had stood a river of oil gushed out. It was a vivid dream but Kepler attached no particular importance to it at the time. Some time later, however, he visited his brother on the Egbert farm and saw there the scene of his dream. He marked the spot where the Indian had stood.

The brothers could lease but one acre of land, and to get that they had to agree to pay a royalty of three-fourths of any oil found. They took in two other men as partners and began putting down a well, which they named the Coquette. The well came in at fifteen hundred barrels a day. In six months it had dropped to six hundred barrels a day, but in those days oil sold by the gallon. The first ten thousand barrels brought ninety thousand dollars. Kepler's dream netted him eighty thousand dollars in six months.

But that was not all. The marvelous dream had made the Coquette a wonder. Every visitor to the oil region wanted to see it. The partners built steps to the well and collected a dime from each spectator.[3]

At the time of his dream Kepler had a job and was getting along as well as most young men. But in 1908 Mrs. Weger of Crawford County, Illinois, was having a hard time. Her husband was dead and she was trying to support herself and children on an eighty-acre farm. She could not afford a hired man, and she and the children had to do all the work. Most of it fell on her, for she sent the children to school when they had shoes and warm enough clothing.

One night she dreamed that she was sitting in a fine room in which there were a piano, upholstered furniture, and a Brussels carpet. She was crocheting, something she liked to do but never had leisure for. She dreamed the dream again. The third time she dreamed it, she, in her dream, got up and walked out of the house. Back of the henhouse she saw a strange contraption moving up and down. She had never seen an oil rig, but later when she did see one, she recognized it as the strange contraption of her dream. So when an operator offered to lease her land, she consented on condition that he write into the lease a clause permitting her to choose the location of the test well.

When the time came to drill, Cramer, the assistant production superintendent, sent the drilling contractor, O'Mara, down to the farm to drive the stake, telling him to ask Mrs. Weger where to drive it. He had divided the farm into eight ten-acre tracts, and if the test brought oil, a well would be drilled on each. When the drilling contractor arrived, Mrs. Weger was not there. Instead of looking her up, he drove the stake on one of the corner blocks, got in his buggy, and drove back to town, about twelve miles. There he told the assistant superintendent what he had done.

Cramer insisted that the terms of the lease be observed to the letter. They went back and found Mrs. Weger and told her to pick the location. She chose the spot back of the henhouse where she had seen the contraption in her dream. The well was a good producer. On six of the remaining tracts good wells were brought in. But the eighth well to be dug, the one on the location O'Mara had made, was dry. If this had been the first well drilled, the lease would have been abandoned, and Mrs. Weger would not have got her fine room, her piano, her upholstered furniture, and her Brussels carpet. Nor would she have had time to sit in her chair and crochet while her children went to school properly dressed.[4]

Mrs. James Rust never knew the poverty from which Mrs. Weger suffered. She and her husband and children lived on an eighty-four-acre farm near Ranger, Texas. One of her sons, John, recalls that during his childhood he heard his parents tell "a countless number of times" of a dream his mother had before he was old enough to take note of such things. His mother woke one morning and said, "Jim, we'll never sell this little farm. Regardless of what happens, we'll hold on to it as long as we live."

He said, "Why, Mary, do you feel that way about it all of a sudden?"

She said, "Last night I had a dream. Look out this kitchen window, up there at the side of the hill, will you, just a hundred feet away? See that old live oak tree out there, the largest tree on the place?"

He said, "Yes, what about it?"

"Well, right there under that tree is where we will find our fortune, because in my dream last night it was very vivid—the picture of that tree—and our fortune will be found right there under that tree. Now in what form my dream did not let me know, but I know . . . that dream was more than a dream—it was a vision."

She repeated the story of the dream many times as John was growing up. In 1915 a neighbor, Jake Baker, who had become famous in that section of the country for his success in locating water wells with a forked peach limb, came to the Rust home and said that he had come to have a feeling that there might be silver up there behind the house. He said he'd like permission to do some prospecting up there. Rust told him to go ahead. He'd give him half of all the silver he found. The next morning Baker cut a forked limb from one of Rust's peach trees and tied a dime to it, and began walking over the farm. The limb turned down under the live oak of Mrs. Rust's dream. There Baker drove a stake and told John that he was under no circumstances to move it. Mr. and Mrs. Rust came out and looked at it.

Baker had an auger made at the blacksmith shop and began boring by hand. It was hard work, but he kept at it until he had gone down a hundred or so feet. Then he lost hope and filled the hole.

A little more than two years later oil was discovered first on the McClesky farm adjoining the Rusts', and a few days later on the farm of another neighbor, Mrs. Nannie Walker. Oilmen rushed in to lease all the available land. Rust made a trade with the Texas and Pacific Coal and Oil Company. In a few days rig builders came out and unloaded timbers and lumber by the big live oak. Then they cut the tree down to make room for the rig.

The well came in, making ten thousand barrels a day. Producing wells were later drilled on other places on the farm, but the dream had not said that the big live oak marked the only spot where a fortune was to be found, and John Rust may not be wrong in concluding:

"And I guess there might be something in dreams. There was something in that dream at least. The fortune was there, just like the dream had told my mother years before."[5]

3

The X-Ray-Eyed

A WRITER who visited western Pennsylvania during the first oil boom in American history reported that "a new class of people has sprung into existence under the cognomen of 'oil smellers,' who profess to be able to ascertain the proper spot by smelling the earth. Some of them practice considerable mummery in order to mystify their employers."[1] These men must have been relatively numerous, for "oil smeller" became a generic term applied metaphorically to all oil locators, including those who relied on hazel bows and second sight.

The oil smeller, however, in the literal sense did not last long. An acute sense of smell such as a few people are endowed with would be valuable in locating the oil and gas seeps which provided the first clues to underground oil, and it is not improbable that some of the early oil smellers made discoveries which may have led to commercial development. But these obvious indications of oil have long been explored, and the oil smeller now exists mainly in the legend of Kemp Morgan, to be examined later.

The second-sighted, the X-ray-eyed, lasted longer. The persistence of a belief in X-ray vision, along with many other folk beliefs, has no doubt been encouraged by the press and the news services.

As late as 1947 the International News Service distributed a story about Peiter Van Jaarrsveld, a sixteen-year-old boy from South Africa, who could "discern the existence of gold and diamonds under the ground." Gold was indicated by a black ridge that appeared before his eyes, diamonds by a "heat blaze," and underground water by moonbeams.[2]

In the 1920's a geologist in Abilene was called upon by a man from a town in an adjoining county seeking employment or a partnership. His peculiar qualification for the position he sought lay in the fact that he had been born with a caul on his head and could see oil under the ground. Not only that, he could tell the depth and give a highly accurate estimate of the production. He couldn't operate on his own hook because people could not understand and therefore did not believe in his mysterious power. What he needed was an association with somebody in whom the people had confidence. The connection would be highly profitable for them both. He did not say why he didn't find oil, lease the land, and go into production business for himself, but he obviously lacked capital. The geologist, with typical skepticism, declined the partnership and has not heard of the man since.[3]

A Negro boy was brought to Ranger in 1917 to locate oil. His white master had fitted him out with a joint of stovepipe, one end of which was trimmed and padded to fit his head in the manner of a stereoscope. The other end would be buried in the ground a few inches. The boy would take his seat on a portable stool and look, or pretend to look, down the pipe. His master would ask, "What do you see?" and the boy would reply, "I don't see nothing." The question would be repeated at intervals and in most instances the boy would eventually say, "I see something dark a-flowing."[4]

I am told that the man did a thriving business. How much of his income went to the boy nobody seems to know.

But when old-timers speak of the X-ray-eyed boy, they are referring to another boy, to another place, and to an earlier time. The usual statement is that a few days after the Lucas Gusher came in at Spindletop in January, 1901, some local capitalists at Uvalde organized a company called the X-Ray Oil Company to drill on locations made by the X-ray-eyed boy, that they brought the boy to Spindletop where he made a location, but that drilling failed to bring oil.

This is not quite as it happened. In 1956 the boy, Guy Findley, at the age of sixty-eight, permitted a tape recording to be made of his recollections of the event and those leading to it. He grew up in the region west of San Antonio, where his father had ranching interests. He had his high-school education at Uvalde High School and went to college at Southwestern University at Georgetown, Texas. As far back as he can remember, he used to "have a feeling" when he went to the spring, a feeling he learned to associate with underground water, and he assumed that when he had that feeling, wherever he was, there was water underneath. But at that time he thought nothing of it. He thought "everybody could do it."

He was about ten years old when he came to realize that his power was extraordinary, and he spoke to his father about it.

"Well, we will try it," his father said. "You show me where there is water." His father had a ranch near Sanderson, a dry country where water is hard to find. Guy found a place where he said water would be found at about sixty feet. His father had a well drilled and got water at sixty-two feet.

"The news spread like wildfire," he recalls, and the boy was in great demand. "Everybody was wanting me to go." He went to Del Rio and located three wells for a man named Stickler. Two of the wells produced water, but not the third. But Stickler had faith in him and suggested that maybe the hole was crooked, that the drill had been deflected so that the bottom of the well was not immediately under the spot Guy had marked. He set off a charge of explosive at the bottom of the well and got plenty of water.

Guy located other wells. He felt that he should not commercialize his gift, but he got to wanting to charge people for it, and did. Those seeking his services would pay his expenses and give his father five hundred dollars if water was found.

He experimented with a peach limb. He would get a clean kind of wood, one that was limber, and take the forking ends in each hand and the other end up, and give it a bend and start walking. If he found water it would start down and he couldn't stop it. It does not work for everybody. He found that it would work for him but he never did use it in locating water. He did not need it.

It was not true that his eyes did not match, that one was blue and the other brown. They are both dark brown. It was not true, either, that he could see under the ground. The term "X-ray-eyed boy" was a creation of the newspaper reporters. But that is what he got to be called until he was grown, and even after. When he was courting the girl he afterward married, she recalls her girl friends spoke to her about him. They said they didn't like for him to look at them. "Don't it make you feel funny?" they asked. "No," she said, "I don't see that he is different from other men."

In fact sight was not involved at all in his locating water. He preferred to work at night, the darker the better. If any image came to his mind, it was not of what was under the earth but some symbolic vision like a spring. But what told him about the water was a hard-to-describe feeling, apparently a physical sensation.

He was thirteen years old and locally famous when the Lucas well came in. It immediately occurred to some men with money to invest that if he could locate oil as successfully as he had water, they could make a fortune.

"Well," Findley recalls, "they wanted to test to see if I could locate oil. So they buried a barrel of oil and a barrel of water out where I didn't know where it was, and took me out there. This I did at night. . . . Well, I located the barrel of oil and the barrel of water. They had them buried about six feet under the ground. So they started out ready to go."

The promoters secured a lease on a tract of land a mile long and about two or three hundred feet wide near Spindletop. They called themselves the Uvalde Oil Company, not the X-Ray Oil Company. Guy's negotiations were carried on by his lawyer brother and he does not know all the details. He made a location on one end of the tract, and according to his instructions told nobody where it was but his brother. Then some misunderstanding arose.

The company refused to sign the contract his brother had prepared and he refused to reveal the location. They drilled anyhow, but at the other end of the tract, and missed the oil. Later drilling on the location Findley had made brought oil. But the lease had changed hands by that time.

During his brief stay near Spindletop, the struggle in his heart between God and Mammon had intensified. He felt that he was abusing his gift, and yet he wanted to get rich. He was not unhappy when his father said, "We'll take him home. I guess we ought not to try to commercialize on it."

He thinks that his gift was impaired by his attempts to make money, though he afterward located some water wells, one for his father-in-law, who had been hauling water every summer when the creek went dry. Even yet, if some man who believes in him comes to get help, Findley will try to find water for him. One man insisted on paying him, but he said no. He believed his gift was given him to help others, and maybe it would come back to him if he didn't charge. He thinks it has.

He is a deeply religious man. He belongs to the Methodist church, though he does not attend services regularly, partly because his dairy farm keeps him at home and partly because he feels "that a man should live his religion," and that he doesn't "have to go to church four or five times a week to make [him] good."

He never prays in public and his private prayers are unspoken. Each day he withdraws for a period of meditation.

I can go out by myself, and everything is quiet and sit down and concentrate, and I don't know what the feeling is, but pretty soon I am through with it. It's over, and I feel being way off into something —communion with the Deity—whatever it might be. I don't know.[5]

Findley had heard of an Egyptian girl who could locate water in much the same manner as he. He had not heard of Augusta Del Pio Lougo, who reported that when she walked over underground water little shocks passed from her feet to her head, "causing distinct pain,"[6] and who found that she could also locate oil; or of Evelyn Penrose, who when walking over oil felt a "violent stab in the soles of my feet like a red-hot knife."[7]

In Texas Augusta and Evelyn might have been called oil jumpers. One man so called came into Ranger during flush times, arriving on the train. His pains seem to have been more severe than those of Augusta, or Evelyn, for as the train neared the station, he began jumping up and down in the aisle and begging the other passengers to shoot him and put him out of his misery. When questioned, he said that when he was over oil, he suffered such exquisite torture that death would be welcome. He got off the train still jumping and could not be at ease until he was taken to a farmhouse some miles from town. He eventually left Ranger richer and perhaps a more cynical man.[8]

A well was subsequently drilled on the farm where he had stayed. It was a dry hole.

An oil jumper is not the same as an oil tromper. The late H. H. Adams, geologist of Abilene, Texas, was at work near a road in the Desdemona area one day when a farmer came by and stopped. Adams, to start conversation, pointed to a leaning chimney on a farmhouse nearby and remarked that it ought to be straightened before it fell down. The farmer explained that the chimney was not really leaning; it just seemed that way. There was so much gas pressure under the ground that the normal processes of gravity were interfered with. Things that were level did not look level, and things that were straight looked crooked.

Asked how he knew about the gas pressure under the ground, the farmer said that his brother was an oil tromper and that he had walked over the farm. If you are an oil tromper, when you walk over oil, you will make tracks twice as deep as the tracks you make when you are not walking over oil. When his brother walked over the farm he sank down twice as far as the farmer did. He was a little heavier but not much.

He hadn't seen his brother for some time now. He kept hoping that he would come to see him soon, but he doubted it. He was too busy tromping oil for Standard Oil. He was just tromping himself to death for them and they wouldn't even let him off to come and see his folks.

Several times during the next few months Adams stopped at the farm hoping to meet the oil tromper. But he never did. The man was always on the job, just tromping his life out for Standard Oil.[9]

Nor have I been able to find him. But it is a fact that when a well was put down on the farm with the chimney that seemed to lean but didn't, oil was actually found.

4

Seers

FROM THE BEGINNINGS of the oil industry until today there have been oil locators whose methods were professedly occult. They get their information from the dead, from the stars, and from whatever other sources fortune-tellers have used through the centuries.

Abram James, who is given credit for the discovery of the Pleasantville, Pennsylvania, field, professed to be guided by the spirit of a Seneca Indian. In the fall of 1867 he and three other men were driving from Pithole to Titusville. As they came to the boundary of a farm belonging to William Potter, James stopped the buggy, shuddered for a moment, and began muttering in what his companions assumed to be the Senecan language. Then he jumped out of the buggy and went running across the field. When he had gone some three or four hundred yards, he stopped, spun like a top, and fell to the ground. When his friends got to him, he had a finger thrust into the soil and was unconscious.

When he revived, he said that his friendly spirit had revealed to him that oil was to be found at the spot marked by his finger. So sure did he feel of the accuracy of this information that he proposed to drill. He could not himself finance a well, so he asked his friends to join him in a partnership. Not sharing his conviction, they declined, but James succeeded in borrowing enough to put down a test well. The well came in in February, 1868. With the help of the Indian spirit, he located other producing wells in the Pleasantville area, but when he ventured elsewhere he found the spirit's advice unreliable. He got only dry holes, but he soon recognized the limitations of his guide and quit sending good money after bad. His production at Pleasantville made him a wealthy man.[1]

It seemed appropriate to call upon a Seneca in western Pennsylvania, for that was the ancestral home of the tribe, and the first oil, gathered from seeps and marketed for medicine, was called Seneca Oil. Across the continent the spirit of another dead Indian proved friendly. The discovery well at South Mountain, California, was located upon the advice of a geologist, but the principal financial backer, a believer in spiritualism, would not advance money for drilling until he had consulted his mentor, the spirit of the dead Indian. The Indian gave his OK, and thus "geology and spiritualism in conjunction with Lady Luck were successful."[2]

The race of the spirit that advised a Louisiana farmer is not revealed. The Gulf Oil Corporation had drilled a dry hole on the farm. Later a new field was opened a few miles to the north, where the farmer also owned land. But he refused to lease unless the company would agree to drill a well on the old lease near a tree, where a spiritualist had told him oil would be found. Efforts to dissuade him failed, and in order to get a lease in the proven area, the company agreed. The well came in at five hundred barrels a day, but tests around it were failures. This was the rare phenomenon of a one-well oil field.[3]

It was inevitable that the professional medium should open shop to give advice to oilmen, speculators in oil stocks, and landowners. The spirit that he—or usually she—called up needed to be friendly to the person seeking counsel. It would seem that it need not be that of one who as a mortal was familiar with the region or with the oil industry. The mediums did not confine their advice to oil matters, or even to business. They drew clients from many occupational groups. But it has been noted that they flourished in the oil regions, particularly in boom times.

In 1919 a newspaperwoman called upon a number of spiritualists practicing in Fort Worth and found them doing a land-office business. She wrote in the *Star-Telegram* for February 9, 1919:

Spiritualism has entered the oil game. Invoking the counsel of the deceased to locate wells is the favorite panacea for the harassed minds of speculators. Mediums are giving engagements weeks ahead to persons who come to consult them on the great question of the hour....

During the past week I visited several mediums who are the most popular. I found their anterooms crowded with both men and and women....

Every person [one medium explained] has his or her guide in the spirit world, some soul that has passed beyond. The souls of loved ones watch over us here on earth, and would make their messages known to us if we could but understand. It is the duty of the medium to be interpreter between the material and spiritual world.

And since a spirit has no body, it can readily enter the interior of the earth and report what is there.

The clients of these mediums were not revealed by name. And not many men have placed themselves on record as having consulted tellers of the future. Among the few who have is George W. Weller. Weller was a young man working in Beaumont for the *Galveston News* and *Dallas News* when the Lucas Gusher came in in January, 1901. He opened a small store which eventually became a large hotel supply company. He did a great deal of trading besides, buying unclaimed freight or anything else, including a shipyard, which he thought he could sell at a profit. And he always could.

Soon after he had opened the store in partnership with his mother, he went to a fortune-teller who called herself Zara. She had recently operated in San Antonio, but the oil boom had brought her to Beaumont, and she was getting all the clients she could see. She had Weller write his name on a piece of paper and fold it into a tight wad. Without unfolding it, she held the paper against her head and told him his name, the date of his birth, and all the places he had lived. She said he was in business with a woman, whom he would soon buy out. Soon afterward he bought his mother's interest. Zara told him that he had a sister who taught music, and that she would soon quit teaching and give her piano away. Before the year was out, his sister quit teaching and gave her piano to her niece.

At that time Weller had a girl in Kentucky whom he hoped to marry. Zara told him he had not yet met the woman he was going to marry. When he met her they would marry in a short time. Soon afterward he saw a girl going down the street and knew that she was the girl for him. They married. Zara was right about this and everything else she told him, including the number of children he would have.

She told him he would be successful in business, that everything he touched would turn to money. Every deal he made would make him money, except in oil. He would lose every dollar he put in oil. And she told the truth. He lost every dollar he put into an oil well. One time it looked as though Zara was wrong. He and some associates drilled three wells near Temple. They got gas, and as they were near a pipeline, had a ready market. But their wells soon failed and they never got their money back.

It might have been different if he had bought into a well at Sour Lake. While the well was drilling he was in Sour Lake and was offered an interest in the well for five thousand dollars. He did not have the cash in the bank, but the fellow offered to take a mortgage on the store. Weller said he would think it over and call the man the next day. He decided, in spite of Zara, to make the deal. But next day when he called by long distance, the man said, "We've hit oil. You couldn't buy in for twenty thousand now."

Weller was sorry that he did not take the offer immediately. It would have made him barrels of money. But he remembered what Zara had told him. She was right about everything else. Everything she told him turned out just as she said it would.

The source of that mysterious power by which Zara could tell a man's past and future by holding his name on a wadded piece of paper to her head Weller did not know or learn.[4]

A reader who gives advice on many subjects including drilling for oil is Ruth Bryan, or Madame Virginia, of Abilene, Texas.

She discovered her talent very early in life. Sometime before her sixth birthday, she told her father that some of his cattle had been stolen. He replied that if she told any more yarns like that he would whip her. But in three or four days he got word that twenty-five or thirty head of cattle bearing his brand had been shipped from Brownwood.

Perhaps that is the reason why some time later, when she was six, he asked for her judgment about the Fry ranch. They were visiting there when he asked her what she thought about the place. She told him that she could see ladders all over it and some machinery she thought was a cotton gin. But years later the ladders turned out to be oil derricks and the gin a refinery.

The first oil well located upon her advice was in the Burk-burnett area. At that time she was operating the Alpine Hotel in Wichita Falls, an establishment patronized largely by oil people. One day an oilman asked her if he had any mail. She told him no, but that he would soon get a message that his son had gotten his leg broken in a gate. In two hours the message came. The man then asked her if she would go with him to locate an oil well. She located several, and they were all producers. She was soon pestered by so many oilmen seeking free advice that she began charging for it. Her standard consultation fee is two dollars.

She says she does not have to be present to tell whether there is oil under a piece of land. If a man is talking to her about his farm or ranch, she can see it all very clearly in her mind. One woman came to her for advice about buying a ranch near Anson. Madame Virginia told her to buy it: there would be an oil field on it sometime. The field was twenty years a-coming, but it is there now for everybody to see.

She had never had a client to complain, but she had made mistakes. It would be blasphemous for any human being to claim perfection. Once her vision of the terrain was inaccurate. On the Iatan Flats what she thought was a valley was hills. She said that there was no oil there, but there was.

How does she know where the oil is? She just feels it. She can get on a train at night and tell when the train passes over oil. She can read cards, teacups, and palms, but it's all one thing. It's the presentments that come to the mind. She thinks the Witch of Endor has been judged too harshly. She was just a seer, such as there have always been. They have been mighty few, but always a few. If she had been a bad woman, God would not have let her call up Samuel, one of his angels.

In 1926 a man came to her for a reading. He said he lived in Tyler, but Madame Virginia knew better. "No, you don't," she said. "You live in a real small town." He admitted that he lived in Hawkins, and she said, "There's going to be an oil field in Hawkins." He bought some lots there and now he has oil wells on them.

She herself bought 160 acres between Hawkins and Red Springs, sight unseen. That is, except in a vision in which she saw an old schoolhouse and some derricks. She later visited the place and saw the schoolhouse. The derricks have not come yet, but she got good money for a lease. For a long time she wondered why oil had not been found. Then the answer came to her. Near the line of her property, the oil flows down like a waterfall, so they will have to drill much deeper, and someday she thinks they will.[5]

Annie Jackson, of Corsicana, Texas, who for many years until her recent death practiced healing and gave advice on many subjects including where to drill for oil, believed that her power came from God, and was portended at her birth. She was the seventh child of her mother.

At the time of her birth her mother was working for a white woman, Mrs. Annie Stroud of Groesbeck. As Annie Jackson told the story, Mrs. Stroud one day said to her mother, "Aunt Marget, is you fixing to born a baby this morning?"

She says, "I don't know."

She says, "Yes, you is, Aunt Marget. I'm going to phone to Mexia and get a doctor for you."

So she phoned to Mexia to get a doctor and the doctor didn't come. She phoned to Groesbeck and got a doctor, and when he got there I was done born.

The doctor come in and says, "I love fresh meat." Says, "Y'all done killed hogs." Says, "Here's a nice hog heart, hog lights and chitlings."

Mrs. Annie says, "No, that's Aunt Marget's baby. That's Auntie Marget's baby."

Says, "Well, I've never seen nothing like that. I'm forty-nine years old—not forty-nine years old, sixty-nine years old—(I'm going to tell it just like he said—like she told me he said). I've waited on both colored and white, and never seen nothing like that in my life."

Says, "Sho nuff."

Says, "Mrs. Stroud, would you mind laying a layer of paper on your dining-room table and put a quilt on that and lay a paper on top of that quilt?"

Says, "I'll do anything for Aunt Marget."

"Now put a sheet on there. Let me examine this thing and see what it is."

So when he examined me, my feet was back that way and my hands down toward my thighs, my head down long my shoulders.

He says, "Look here, she's got a mouth full of teeth." Says, "Aunt Marget," says, "This is a clairvoyan."

Says, "What is a clairvoyan, doctor?"

Says, "Some of them calls them fortune-tellers. She's no fortune teller." Says, "She can tell you things under the earth just like she can on top."

Mama says, "I don't want that thing. I don't want that thing a tall."

So Mrs. Annie says, "Well, I'll raise her"; says, "I'll raise her."

A Mrs. Eubank also offered to take care of her. So she was raised "mostly by white folks." There was no colored school nearer than nine miles, so she did not attend. She did, however, worship in a white church, and she is glad she had a religious life. God converted her soul when she was ten years old.

Mama sent me to the field one morning. She says, "Annie, you ten years old this morning. I want you to do down yonder in that field and chop cotton—get a piece of hoe and chop cotton."

So I started to go then, to pick up the hoe and go to the field, and I heard a singing, "Amazing Grace, How Sweet the Sound," and the Spirit said, "Annie go and pray." I dropped that hoe down. I said, "Lord, I can't pray. Lord, I can't pray," and walked right on back up there, and He said, "Annie go back and pray." So I went back down and says, "Lord, have mercy; Lord, have mercy." I jumped right up again, the Spirit followed me right on back. "Annie go back and pray." So I went back down a third time, saying, "Lord, have mercy; Lord, have mercy." So I prayed from May to September. "Have mercy, Holy Ghost."

In September she found peace with God and demonstrated her gift of prophecy. One evening her mother came into the kitchen and said,

"I want y'all children to pick out the dead cotton." (We done picked our best cotton.) "I want y'all children in the morning to git your sacks, and Annie, you go get your flour sack." (You know I was small)—"And I want y'all to go to the field in the morning up here in front of the house."

I said to the children—I was already happy. I says "Um-hum un." I said, "We ain't going to pick no cotton in the morning." I says, "Icicles is going to be hanging like that, tree limbs all broke down in the yard."

The next morning her mother looked out the kitchen door and found the ice just as Annie had said. She said, "I never seen nothing like that before. That child said that last night. Mama ain't going to whip her no more. I started to slap her but I ain't going to slap her. Lookey here. I've seen hard hail. Tree limbs in every yard. Come here children and look."

The children all ran and looked. Annie was sitting in the back room.

... setting down on a little box, just a rocking; Jesus spoke salvation to my soul. "Annie, I want you to give up Mother and Father, brother and sister, friends and relations and give up the world. I'll bring your soul from sin to salvation." . . . I was so glad I didn't know what to do.

She wanted to join the church, but her mother thought she was too young. Annie asked the white folks if she could join their church. They said that she could come and sing and pray with them. So when she could she walked eight miles to the Red Chapel until she moved to Corsicana.

She told the future and the past. "That's all people got, in this world." She read palms, not by line but by the "blood circle." Although she had never been to school a day in her life, she read cards, just as the doctor said she would.

Her first prediction of oil was made some time before 1920. Her brother-in-law, Walt Hosley, had bought, evidently largely on credit, a farm afterward acquired by Julius Desenberg.

He was buying it, and he lost his wife. He lost his wife, and he said, "Sister Annie, I believe I'll give this piece of property back." And I says, "Let me look at you and see if it do any good to give it back." So I looked at his hand and described oil and I says, "Boy, keep it." I says, "It'll pay for itself this time in a year or a year and a half. It'll pay for itself. Here oil."

[He] says, "I thought you had some sense. No such a thing as oil under the earth."

I says, "All right, when you give it up, it's going to be some white man come in here and I'm going to tell him about that oil." Sure enough I told Mr. Humphreys.* I says, "Here's the first well. I told

---

*Colonel A. E. Humphrey brought in the discovery well at Mexia in November, 1920. According to Boyce House the Negroes regarded him as a demigod. "When the Henry—the first of his big wells—was being drilled in, two ex-slaves led the singing of 'Let the Light from the Lighthouse Shine on Me.' When the oil began to flow the crowd took up the song, changing the words to 'Let the Oil from the Henry Spray on Me.' "

The Colonel built a town clubhouse on the Confederate Reunion grounds. "Humphrey's superintendent," writes House, "had difficulty completing the work . . . because during his absences the colonel was likely to appear and call a halt to the work so the one hundred Negroes could sing for him." (*Oil Boom*, p. 137.)

my brother-in-law about this." I says, "You are going to get to scrambling and get the first oil well." So there's the first oil well right there. I told him about it and he got the oil. My brother's gone in now [i.e., has died].

She says that Colonel Humphrey gave her $8,500 which she used to build a house in Corsicana.

She claimed to have predicted oil at many other places, and oilmen came to see her from as far away as California. When William A. Owens called to tape-record the interview upon which this sketch is largely based, he found the waiting room crowded with clients white and colored, and he was unable to see her until next day. She generally stopped reading at dark but sometimes she went on until nine or ten o'clock. She made no charge for healing—it would be illegal for her to—but accepted gifts for it. Her usual fee for other services was five dollars. One white man who consulted her about his income tax gave her ten, and suggested that she should raise her fee. Another man came and said he had five oil wells and was going to give her some money. She told him not to worry about that. "When you go beyond, them that got it lose it; them that don't have it can't lose it."[6]

Much of her money went to religious causes. She said she had built seven churches, and paid for a cottage for colored people at Mexia. She preached on occasion, sang, and composed spirituals. Here is one of the songs that "just come" to her:

> No, you can't do nothing without the Lord,
> Says, you can't do nothing without the Lord.
> Everything you say, Lord, all you do,
> Well, you can't do nothing without the Lord.
> Yea, the Lord, the Lord,
> You can't do nothing without the Lord.
> Everything you say, Lord, all you do,
> You can't do nothing without the Lord.
> Praise Jesus.

And you can't do nothing without the Lord.
He created the world, and He made everything,
And you can't do nothing without the Lord.
Yes, Jesus.
And you can't do nothing without the Lord.
He created the world, and He made everything.
And you can't do nothing without the Lord.
Yes, the Lord, the Lord.
Well, you can't do nothing without the Lord.
Yes, the Lord, the Lord.
Well, you can't do nothing without the Lord.
Everything we say, Lord, all we do,
Well, you can't do nothing without the Lord.
Praise Jesus.
And you can't do nothing without the Lord
And you can't do nothing without the Lord.
He created the world, and he made everything,
And you can't do nothing without the Lord.

Annie Jackson had no record of the number of her clients seeking advice about oil. And how many oilmen consult seers and how seriously those who consult them take their advice, I have no way of knowing. An oilman tells that two of his fraternity in San Antonio do patronize a local seer. When a third man ridiculed their credulity, they said that he did not know the seer. If he knew her, he would believe in her. They invited him to go with them to her place of business. He came away unconvinced of her ability to give valid advice about oil, but he paid her money to tell him which football teams to bet on the following Saturday.

5

Doodlebugs

IT WAS INEVITABLE that the centuries-old divining rod should be used in the search for oil and that many of the searchers should be amateurs looking for oil on their own land. A neighbor of my father's walked over his pastures with a peach crotch and then spent the next ten years grieving about the millions of dollars which underlay his grass but which he could induce no oil company to bring to the surface. Eventually oil was found on his land, but whether on locations he had made I do not know.

Another ranchman whose rod proved at least 50 per cent right was George Ray, who figures in a chapter of J. Frank Dobie's *Coronado's Children*. Ray was a ranchman and banker who owned stock in a silver mine in Mexico. Once when he visited the mine he found that the men had lost the vein and were having trouble in finding it. Recalling that when he was a boy an old Scotsman had taught him how to witch for water, he cut a switch; and, being of the school that baits the rod with the thing sought, he fixed a silver peso in a split in the end. In five minutes he had located the vein of silver. Later, in 1924, he decided to find out whether he had oil under his land. He cut a peach switch, to the end of which he fastened an "elongated rubber sack" filled with fresh crude oil. Thus equipped, he rode over his range in an automobile. He reported that he had located two of the deepest and strongest pools in Texas. Six years later an oil company brought in a field on one of these locations.[1]

But among the professional oil-seekers the divining rod was soon supplanted by the doodlebug just as the buckboard was supplanted by the automobile.

Doodlebug is the name given to all instruments for locating oil, as well as to the men who operate them. In its simplest form the doodlebug is an adaptation of the baited divining rod. It consists of a bottle or other receptacle containing oil or chemicals derived from oil. This container is covered, usually with a chamois skin. It is suspended by a string held in the operator's hand. It indicates the presence of oil by rotating so that the string describes a cone. The number of revolutions indicates the depth of the oil. One practitioner allows a hundred feet for each revolution. If oil is to be found at two depths, the doodlebug stops and reverses itself, and the number of revolutions made in the direction contrary to the first shows the number of feet from the first oil-bearing sand to the second.

A detailed account of the operation of one type of simple doodlebug has been furnished by a retired businessman in West Texas, whose conclusions are at least disinterested, since he has never located oil for money. He scoffs at the idea that some people are especially gifted. Early attempts to find water, salt water, and oil, he says, were based on superstition. Many people felt forces they could not understand and thus thought themselves especially gifted. But any normal person can learn to witch. Water and oil in their natural states within the earth set up electromagnetic waves, which he calls "sine waves"; these waves any normal person can learn to intercept. But a living person is required. No machine will work without the interposition of a human body. Simple instruments like green twigs, copper rods, or steel wire help. Each individual should experiment with different substances until he finds the one that works best for him. In experimentation, he must go into an actual field. Oil or water in containers will not do, for the sine waves are a result of pressures in a state of nature. The West Texas businessman reports that in his own experience, iron reacts best for water, while nonferrous metals are more effective in locating oil.

Small amounts of these metals either singly or in combination

are placed in a leather bag attached to a string. The presence of oil
will cause the bag to rotate. Many doodlebuggers make the mistake of assuming that the point of highest rotation is the best
place to drill. On the contrary, it will usually be off the producing
structure. The waves come from the ground in conical form just
as light waves from a source on the ground spread out into the
sky, and therefore are felt on the periphery of the oil sand. To
find the pool requires traversing the field carefully and slowly and
marking intervals where radiation is felt, these usually being every
two or three steps. Near an oil pool the pattern will be a two and
three combination. When the center of the pool is reached, the
direction of rotation will be reversed.

To find the depth, the operator will go outside the band of
activity in a straight line, and he will note rotation at regular
intervals, from which he can make his calculations.[2]

Many professionals, however, rejected these simple forms of
the doodlebug in favor of more impressive, if not more efficacious,
devices. Some required batteries strapped to the operator's person.
Some sounded horns or rang bells when they were carried over
oil pools. One pair of doodlebug men, who operated in the Smackover field, had a sedan-chair-like affair borne by two men. At the
beginning of the survey they wound it up and it began to tick.
When they reached the end, they turned a crank and out came a
card upon which was printed the extent of the pool, the depth,
the daily yield in barrels, and the specific gravity of the oil.[3] Other
machines filled whole trucks and had as many dials as the instrument panel of a jet airliner.

The men who operated these doodlebugs, like the vendors of
the nostrums that flood the country, knew the appeal of the word
science to the American mind. To laymen they made pretense to
great learning, acquired before World War II in Germany, then
assumed to be more advanced in science than the United States.
When they tried to sell their gadgets or their services to L. W. Blau,
formerly research chief of the Humble Company, they professed

to be only inspired laymen who had labored many years to perfect their inventions.[4] They were proud that the university scientists could not understand their instruments. University training, they said, built up superiority complexes and attitudes of skepticism, and imparted fixed beliefs that certain things could not be done. Brilliant laymen, therefore, must perform the miracles which the orthodox scientists think impossible.

But each propounded a "scientific" principle upon which his gadget depended. One man after long research had discovered that all matter had sex characteristics and that the sex principle of oil was female. This discovery sent him on another long quest to find the corresponding male principle. At last he had found it and had it with him in a jug.

"There is no scientific reason," he explained, "why a breath, a vapor, an effluvium of petroleum should not rise through the porosity of the earth and report itself to the sensitive chemical sympathy of my discovery."

A man in Fort Worth invented a radioscope which he said (in offering to sell stock in his concern) reacted to the ethereal waves set up by the collision of atoms in oil, and since these waves were of a different frequency from those of other liquids in the ground, the radioscope could detect them.[5] Another doodlebug made use of especially sensitized photographic plates. They were contained in a holder through which the radiations of oil could not pass. Nevertheless the radiations of oil did pass through the earth and register themselves on the plate when it was removed from the holder and held horizontally over the ground between twelve and two o'clock in the daytime.

Upon one of the devices brought to him, Blau reports:

The most versatile radiation sensitive device consisted of a black rubber rod about six inches long on which was mounted a ball bearing. A brass rod carrying an adjustable weight on one end and a removable capsule about two inches long and one-fourth inch in diameter at the other end was fastened to the ball bearing at right angles to the rubber handle.

When the handle was held in a vertical position the brass rod could rotate in a horizontal plane. The radiations from oil were said to issue from the ground in helical paths thus causing the rotation of the movable system. The speed of rotation was indicative of the gravity of the oil and also, in some involved manner, of the depth.

When looking for minerals, it was necessary to remove the oil capsule and substitute for it one which would respond to the particular mineral. For really accurate work there were additional capsules which could be used to determine fine differences in the gravity of oil.

Blau never refused to examine any doodlebug brought to him, and always gave the inventor an opportunity to prove its worth. The first test was to find a big can of oil which Blau kept somewhere on the three floors occupied by his department. A doodlebugger either refused to take the test or undertook it and failed. In either case he had an explanation ready. His instrument would be, or had been, affected by the vast tonnage of steel in the great office building. In such a building full of people the quantity of lipstick, hair oil, and other oleaginous substances set up counter attractions and made accurate recording impossible.

If he persisted he might be permitted to make a field demonstration. Here is Blau's report on such a test given the instrument just described:

It was exceedingly interesting and gratifying to see the device start rotating on approaching a producing well. Only the inventor could hold it, and it was necessary for him to be in motion to receive an indication, either walking or riding in an automobile. A few times, when we "chanced" rather suddenly upon a producer hidden in the timber, we observed violent rotations while a salt water capsule was being used, but who can say that there was no salt water below the oil? In every instance the device indicated oil after the oil capsule was inserted.

The device was further demonstrated on an oil tank farm. It gave rapid rotation on a tank containing ten feet of Conroe crude which had originally been supposed to be full. The explanation was that the gravity of the oil was responsible. In order to test this explanation further, the inventor was asked to try a tank of Sugarland oil. He did not know that the tank was empty at the time. The device rotated faster than ever, supposedly due to the gravity of the Sugarland oil which should have been in the tank.

Two other men had a versatile instrument which would have been invaluable to enforcement officers during the era of prohibition.

Holding the instrument vertically with both hands on the handle, it was found that the baited swinging end came to rest in the direction of the greater oil fields; thus it was possible to point to Conroe, East Texas, Sugarland and others from a room in a Houston hotel. When the bait was changed to whisky, the device in the hands of the inventor stubbornly pointed to a leather bag lying on the bed. The inventor asked his friend how this could possibly happen, since they had finished the last bottle that morning and had not bought more. Upon opening the bag a pint bottle was revealed and the friend admitted having bought it that afternoon without telling the inventor about it. Thus it was proved that the device was not manipulated by the operator.

The successful doodlebugger who sold his services to oil producers never went into a field without considerable detailed information about the region.

The late M. G. Cheney, geologist, then of Coleman, Texas, was employed by a utility company that furnished gas for several West Texas towns to make a location for a gas well. After the rig had been set up, a doodlebug man, whom, because of his mustache, Cheney remembered as Old Handlebar, came along and persuaded the engineers in charge to allow him to make a location. His instrument used bait, of which he had forty kinds—iron, gold, silver, gas, oil, and so forth. He had got the attention of the engineers by going to a certain spot near an abandoned well and reporting the presence of iron at a certain exact depth. They looked up the log of the well and found that at the exact depth specified a bailer had been lost. Although the log did not show it, Cheney knew that the bailer had been recovered.

On the strength of Handlebar's prognostication, the engineers ordered the rig moved at the cost of four or five hundred dollars. The well produced two barrels of oil a day, but not so much as one cubic foot of gas. Handlebar discovered that he had made a slight mistake. He had carelessly baited his doodlebug with oil instead of gas.[6]

H. H. Adams believed that all doodlebuggers were conscious frauds, and he took great delight in exposing them. Adams was employed for two years by the Department of Justice to investigate the geological claims set forth in the promotional literature of Frederick A. Cook, and his testimony was an important factor in the doctor's conviction of using the mails to defraud. Because of the reputation thus gained, Adams was frequently called on for expert opinion on doodlebugs.[7]

A group of bankers at Graham, Texas, had, upon the recommendation of a doodlebug man, put up $32,000 for a block of leases. Some man among them had insisted upon writing into the contract a clause which permitted them to place the money in escrow until the location could be checked by a geologist; and after the first excitement had died down, they were all eager for the check. They employed Adams to make it. The doodlebug man wore a belt from which protruded small coils of copper wire. At first he refused to explain the principle of his instrument, saying that it was a trade secret. But Adams pressed him until he gave an explanation. He said that radioactive waves emanated from the center of the earth. There were a few people, of whom he was one, sensitive enough to feel them. Since oil is an insulator, the waves would be absent over an oil pool. The coils amplified the waves somewhat, but not enough for ordinary men to feel them. Upon being questioned, he emphatically declared that sight was in no way involved.

After some further argument Adams obtained his consent to make some locations while blindfolded. Adams and his assistant, Harry Hotchkins, led the blindfolded man over the area. He made thirty-two locations. At each Hotchkins filed under a small pile of rocks a numbered card with the locator's findings written on it. They led the man around until they knew he was thoroughly confused as to directions. Then they led him back to the locations already made. On the second trial he reported unfavorably on twenty-seven of the thirty-two. When this fact was revealed to him he seemed greatly aggrieved to think that one charlatan had betrayed another.

Adams used a similar stratagem to expose another man, who had made locations for an operator in East Texas. The doodlebug man had driven a stake in the corner of a cotton field bordered on two sides by pine forest. Adams led him through the brush until he was completely turned around. As they struggled through the thickets Adams would ask him about the probability of finding oil at various places. They eventually approached the man's original stake through the woods. When they were only a few feet from it, Adams remarked that the place looked good to him, whereupon the man assured him that there was no oil near. When he saw the stake, he wanted to fight.

The cleverest fraud that Adams ever exposed was a man whom an operator, an official of a large company, had brought down from New York. The man claimed to be an Englishman and talked like one. He said that he had been a member of the Royal Geological Survey for India, and that the Survey had published several of his papers. He gave titles of articles and numbers of bulletins.

He said that while on the staff of the Royal Survey he had made the location of water his specialty. He had perfected an electrical device which had been completely successful. Sometimes, however, the machine, instead of behaving as was to be expected, would sputter and crackle in a most surprising manner. He noted the places where this had happened and eventually prevailed upon some British capitalists—purely in the interests of science—to drill on one of them. The result was the discovery of the Burma oil fields.

He then proposed to use the device commercially, but the Survey forbade him, on the ground that since he had perfected it while in the service of the state it belonged to the state. He then resigned and managed to smuggle his machine into the United States.

It was the most impressive device that Adams had ever seen. It was mounted on a tripod and operated by an automobile battery. It had five or six dials and various coils and condensers. It was never set up in the open. Before it was assembled, a tent would be stretched to house it. The precaution was absolutely necessary. The instrument was so sensitive that the sun's rays would set it on fire.

The man had evidently studied the field, for on structures not known as proven territory the dials would behave nicely. But over charted pools the machine would pop and sputter and shoot out sparks in a most frightening manner.

Adams knew nothing to do but to watch the operator carefully. He observed that when the sparks flew the man always had his hand on some one of the legs of the tripod. Further observation showed that there was a copper wire imbedded in the wood of each leg, around which there was a metal ring. The operator would slip the ring down until it made contact with the wire.

But he was so skilful that it required three days for Adams to note this. Before the beginning of the field test, Adams had had his employer cable the Geological Survey of India. An answer had come that among their publications were no such titles as those the man claimed to have written; nor was the pretended author's name known to them. The producer was present with the cablegram in his pocket when the final test was to be made. Adams asked if he might see if he could operate the machine. He reached down and moved the ring. The machine sputtered and the sparks flew. The producer read the cablegram. He told the imposter that he would send him to prison except that he did not want to appear in court as the jackass he had been, but that if the man were not out of the country in twenty-four hours, he would have him arrested. The man left.

The most legendary as well as the most successful of the doodle-bug men was Dr. P. S. Griffith. Soon after the Lucas well came in he appeared in Spindletop and for years operated in the Gulf Coast region as well as in California and Mexico. He had practiced medicine in Mississippi, and in the oil fields maintained the dignity displayed by professional men at the end of the century. He always wore a stovepipe or derby hat and a black Prince Albert coat. His beard was long and black. His mien made his predicament comical to those who saw him inadvertently step into a slush pit at Spindle-top and sink in mud almost to the neck. He was able to laugh about the incident, then and years later when W. M. Hudson saw him in the Tampico field in Mexico.[8]

Presumably the Doctor had developed his instrument while practicing medicine, and by 1901 had sufficient confidence in it to abandon his practice and move to the new oil fields in Texas. What was in the box that housed the chief unit of the instrument, he did not reveal. From this box protruded three flexible tubes, one terminating in a bit or plate to fit the mouth, the other two in small handles. In front was another tube or coil fitted on the end with threads for the attachment of metal capsules about the diameter of a pencil and about three inches in length. He had a dozen or more of these, labeled *oil, gas, saltwater, gold, silver, sulphur,* etc. In prospecting for oil, he would attach the appropriate capsule, take the bit in his mouth and a handle in each hand, and start walking over the land. If he passed over an oil-bearing forma-tion, the capsule would turn down. He would mark the spot and walk along a line at right angles to the first. By following this gridiron pattern he would map the field. He said frankly that he could not tell the depth, but he thought he could determine the lateral limits of the oil very accurately. He would sometimes advise moving a location only six or eight feet. He once told Al Hamill that if Hamill would relocate across the fence, a distance of not more than 150 feet, he would be happy to put some money in the well. Hamill completed the well on the original location and it proved a good producer.[9]

Griffith made a favorable impression on Hamill's associates by telling them what had happened to their well. This was in California. A slight earthquake had caused the slipping of underground strata. The casing had been cut in two. The lower part of the well was some feet out of alignment with the upper part. When they resumed drilling, therefore, they had two holes below the point of slippage. This, presumably, was unknown to the Doctor, who walked around the location with his instrument and told them what had happened.

His instrument was not infallible. He explored the region along the Balcones Fault in Limestone County, Texas, and pronounced it dry. In the 1920's several large fields, including Mexia, were discovered there. But his success in the early shallow field at Humble was conspicuous and made him and his partner wealthy. The Doctor's wealth, however, was considerably reduced by subsequent failures. He explored the region between Humble and Baytown and pronounced it rich, but drilling failed to produce much oil. In the long run, too late for him to benefit, he was proved right. The oil was there, but it was too deep to be reached by the drilling equipment then in existence.

One man who was associated with some of Griffith's clients watched him work carefully and came to the conclusion that his instrument was for show only and that his locations were based on surface observations.[10]

The doodlebug flourished especially in the boom time of the past, when it took a man of more than common stability to maintain his judgment in the feverish excitement of an oil boom. When wildcat territory was being drilled, when dry wells were being put down on some tracts and gushers were being brought in on others that looked no more promising, and when men were frantic to know where to drill or where to invest, and no man of science professed to tell them with certainty, they would listen to the man brazen enough to speak with conviction.

Some wildcatters, however, have employed doodlebug men with the same deliberation with which others have employed geologists and geophysicists. O. W. Killam, who discovered the Mirando and Schott fields near Laredo and who has been called the Kit Carson of oil, is one of these. He and his associates made their decision to acquire acreage in the region on the basis of the similarity of the terrain to that of the Bartlesville field in Oklahoma. A friend of Killam's who had been trying to perfect a doodlebug which worked on chemical principles offered to prospect the land for expenses during his vacation. A high percentage of his locations proved productive.

Later Killam sent another man out to look over an area. The man returned and said, "Well, I bought you a lease." He said the country looked good, but he could find only one lease for sale and he had bought it. Killam replied that the only reason the lease was for sale was that nobody wanted it. He told his lawyers to look for a defect in the title so he could turn it down. The title, however, was good and Killam paid $6,000. That lease has produced $5,000,000 in oil, and in 1956 was still bringing in about $30,000 a month. It produces from four different sands and has not been fully developed.

Killam did not recommend doodlebugging. He merely stated that it had worked for him better than anything else. He observed that nobody, whatever his technique, can say with certainty that oil will be found on any given location. The wildcatter cannot drill at random. He has to have some reason for believing in the probability of success. Somebody has to give him faith in some specific location.[11] This is the common function of the geologist and the doodlebug man.

6

Oil on the Wastelands

SOME TIME AGO a woman of my acquaintance received a letter from an oilman wishing to lease a few acres of land she owned in East Texas. She had hardly thought of the land as hers. She had received it from her father, who had thought so little of it that he had not bothered to pay the taxes on it for more than twenty years. Now in order to perfect the title and make a valid lease she would have to pay in back taxes, penalties, and interest far more than the land was worth, or at least far more than the surface was worth.

She asked the advice of her banker, a man as conservative as most of his profession.

"You say the land's no good?" he asked. "Then by all means perfect your title. They'll be sure to get oil on it."

The lease is still in force, but no well has been drilled and none is in the drilling at present. It is too early to know whether the banker was right or wrong. There was, however, nothing original in his prediction. He was merely restating a belief widely held in the oil regions. "There must be something under the land, for there is surely nothing on it," one hears; or, "Since God made nothing useless, He must have put something under this land."

But the argument is usually empirical rather than theological, and is as old as the oil industry in the United States. It arose with the first discoveries in the rocky hills of Pennsylvania and West Virginia, and was given further impetus by subsequent discoveries in the swamps of Ohio, the saltgrass flats of the Gulf Coastal Plain, the arid regions of the West, and in many other places hostile to farming. It is not that oil has never been found on fertile land;

it is that one cannot reckon the true value of land by a mere look at the surface—a fact which has in some instances added to the history of the oil industry a spice of ironic justice.

Consider the Englishman who disapproved of the democratic tendencies of his daughter. We will call him Dorset because he claimed to be of a noble family although not himself a peer. He brought his family to live in Kansas, where he had invested heavily in land and where he expected to live the life of a country squire. One day after they had been living there for some time, he met his daughter on the street, and in the presence of others angrily demanded to know if what he had just heard was true—that she had been keeping company with a common farmhand. She replied that she had danced several times with a young man who worked on a farm. Dorset shouted that if he ever heard of her being with this man again he would disinherit her.

She not only saw him again soon—she married him. They went to Indiana, where the young man, whom we will call Jones, had come from, and with his three-hundred-dollar savings and a loan they bought a farm not far from Indianapolis. He had been able to borrow the money for only eighteen months, and by the time he had harvested his first crop, he knew that unless he could refinance his mortgage he would lose his farm. Moreover, he had found the working of it a backbreaking task, for it proved to be heavily infested with sassafras.

In the meantime Dorset had divided his property among his children. He decided not to disinherit his daughter after all. He thought of something better. He gave rich valley farms to his sons, but to his daughter he gave a barren, rocky hill which would surely keep her poor the rest of her life if her husband tried to make a living on it.

Back in Indiana one day the farmers of the community had gathered to help each other dig sassafras roots from their fields. They wondered what had happened to Jones. He was a good neighbor and a good worker, but he hadn't shown up. About midmorning he appeared all dressed up in his Sunday clothes and with a telegram in his pocket.

"Do you think it is Sunday?" one of the neighbors asked. "Your calendar must be eight years old."

"No," said Jones, "I'm going to Kansas."

"Are you crazy?" the neighbor asked. "You know you haven't any money to waste on a trip to Kansas."

Jones displayed his telegram. "I have here an offer of twelve thousand dollars for my wife's land."

"Don't wait for the train. Wire," a neighbor advised him.

But Jones said no. He and his wife took the next train. In a few days they returned with twenty-five thousand dollars bonus and lease money. Later wells were drilled and their royalties brought them much more.

But the brothers to whom the father had given rich valley lands had only their farms, for all the wells drilled on their land were dry.

Consider too the case of Jackson Barnett.

When under the Dawes Act of 1887 and the Curtis Act of 1898, the United States government moved to destroy the tribal government and land system of the Creek Indians, the majority acquiesced, more in submission to the inevitable than in a joyous acceptance of their new rights and responsibilities. But a minority resisted. Their leader, Crazy Snake, carried in his pocket a printed copy of the treaty negotiated with his people before their removal from Georgia, and he had a disconcerting habit of reading from it, even to congressmen and senators: "The Creek Country west of the Mississippi shall be solemnly guaranteed to the Creek Indians, nor shall any State or Territory ever have a right to pass laws for the government of said Indians, but they shall be allowed to govern themselves." But the white man's answer was always the same: the treaty had been superseded by laws passed by Congress, and the tribal lands must be divided into 160-acre tracts and allotted to individuals.

Crazy Snake's faction organized a tribal government and passed laws forbidding Creeks to accept the land. He and some sixty of his followers were arrested and put in jail. They were released upon their promise to quit trying to re-establish tribal government, but a number continued their resistance passively by refusing to have anything to do with the allotment of land. Their names were on the rolls but they would not make their selections and get their deeds of title. The government officials decided that they couldn't make the Indians claim the land but that they could make them the legal owners. They assigned to each Indian a specific plot whether he wanted it or not. It is hardly coincidental that these plots were on the poorest land of the reservation. And it was hardly intentional that this land would in a few years make some of the followers of Crazy Snake the richest Indians in the entire United States.

The richest of all was Jackson Barnett, an uneducated man of simple mentality and simple wants, who probably had no understanding of the issues involved other than that the ways of his forefathers appealed to him. He did not believe in the private ownership of land and he declared that the land given him was fit only for coyotes and jackrabbits to starve on. He would have nothing to do with it. Instead he lived in a hut near Henryetta, some forty miles away, earning from fifty to seventy-five cents a day working as a farmhand and as a helper on a ferryboat. He was satisfied with his income. It kept him in tobacco—he did not care what kind just so it was tobacco—and enabled him to save a hundred dollars from 1896 to 1912.

Then one day a white man came to see Barnett about leasing his land. This was the first news the Indian had about the discovery of oil at Cushing. The man offered him eight hundred dollars for the right to drill for oil on his allotment and promised him one-eighth of all the oil found. Jackson Barnett agreed. He took the money, and because he could not write his name he signed the lease with the imprint of his thumb. The validity of the document was immediately attacked by a rival leasing agent. A court declared it valid but found Barnett incompetent and appointed a guardian for him.

Soon wells were producing oil on the land which would not support coyotes and jackrabbits, and Barnett's income was more than he could count. But this made little difference in the habits of Jackson Barnett. He no longer had to work on farm or ferry; he had ponies to ride; his guardian built him a better house, a small cottage; and he began buying five-cent cigars. He drew a hundred dollars a month for himself but he rarely spent more than fifty or sixty. The rest was deposited in nearby banks. This peaceful existence might have gone on for a long time had not his guardian been puzzled about what to do with the fortune. By June, 1917, he had on deposit for Barnett about $800,000. He wished to invest it in Liberty Bonds, but having some doubts about the legality of the procedure, he referred the question to Washington. The story got into the papers. In an editorial that Thoreau would have applauded, the *New York Times* wondered why Barnett needed a guardian anyhow.

It is not crime—and it might even be called a virtue that he sleeps on the porch of the little cabin for which he reluctantly abandoned the ancestral teepee. And even if it be true, as asserted, that about $50 supplies all his wants, weakness of mind is not thereby necessarily proved.

A senator introduced a resolution giving the guardian the authority he sought. And Jackson Barnett's peace was at an end. He began receiving letters by the bushel but they did not worry him much, for he could not read them and did not bother to have them read. The world did not make a beaten path to his door, but certain of its inhabitants did get there.

One of these was Anna Laura Lowe, a thirty-nine-year-old widow, who on a day in January, 1920, arrived in a taxicab, with the object, she later swore, of trading in leases. When she left, the seventy-year-old Barnett got in the taxi with her, and she drove him to Okemah, where she tried to get a marriage license. Refused, she drove on to Holdenville, with the same result. She took him home, but three weeks later she came back and took him to Coffeyville, Kansas, where a marriage ceremony was performed. Then they drove into Missouri and were married again. Whether or not she kidnapped him later became an issue between her and the government.

She dressed him in good clothes, had his teeth fixed, took him to Los Angeles, and installed him in a colonial mansion. There he spent his time riding in his cars—he had two—fishing, and unofficially directing traffic on his block.

That is, what time he was not in court. Courts at various times declared him competent and incompetent. He had several guardians. His allotment eventually produced twenty-four million dollars' worth of oil, and Jackson Barnett became widely known as the richest Indian in the world. He gave large sums of money to his wife and to various charities. When approached, he would say that all he wanted was enough to eat on till he died, and being assured that plenty would remain, he would imprint his thumb on the document. Charges and countercharges were made. The attorney general appointed a special assistant to look into his affairs. A series of suits eventually resulted in the cancellation of all his gifts, and, three weeks before he died in 1934, in the nullification of his marriage. What was left of his estate was divided among thirty-four heirs, all Indians except one, the white widow of an Indian. Anna Laura Lowe defied eviction and was driven from the Los Angeles house by tear gas.

Thus Jackson Barnett, the disciple of Crazy Snake, who resisted allotment, became the richest Indian in the United States, thus the residue of his estate was returned to his people, and thus the government that had despoiled him in the nineteenth century made what amends it could in the twentieth.

Though Barnett's fortune was greater than any other, he was

only one of the dissenters made wealthy by the allotments they opposed. But only a minority of Creeks profited from oil.

This was not true of the Osages. This tribe, after suffering more than most from the white man's land hunger, attained the highest per capita wealth of any socially organized group in the world.

This was not due to the benevolence of the white man. When he first came in contact with the Osages, they controlled a large area, including most of the territory now making up Arkansas, Kansas, Oklahoma, and Missouri. By a series of frequently violated treaties and by legislative enactments imposed upon them, their domain was reduced to 1,470,743 acres of poor land in the Cherokee Outlet. One treaty in 1865 deprived them of 1,500 square miles in Kansas and allowed them a compensation of $3,000—two dollars a section. This treaty and another in 1868 were so repugnant to the interests of the Indians that humanitarians denounced them and they failed to secure the two-thirds Senate majority necessary for their ratification. Resort was had to congressional resolution, which requires only a bare majority. It was by the terms of this resolution that the title to Osage lands in Kansas was extinguished and they were moved to a location in the Cherokee Outlet west of the ninety-sixth meridian. The Osages had been principally hunters and traders. Their trade with the Comanches had been destroyed, white settlers were encroaching on their lands, and white hunters were slaughtering the buffalo on their range. They were becoming more and more dependent upon government rations; they had no choice but to submit.

After they had moved and were learning to farm, it was discovered that the ninety-sixth meridian had been erroneously surveyed. They were forced to move farther west and abandon their improvements. From their new location they continued their buffalo hunts with decreasing success until 1876, when all the buffalo were gone and they were again dependent upon government rations. Thus one of the proudest of the plains tribes was reduced to abject poverty.

They did, however, maintain their tribal integrity. Unable to prevent individual allotment of land, they secured the passage of an exceptional provision, renewed from time to time, reserving the mineral rights to the tribe. This was in keeping with their tradition that what a man worked for was his but what came without working belonged to the group.

Aided by government rations and by some income from grazing leases, they eked out an existence on their barren farms. Then on March 16, 1896, James Bigheart, the Principal Chief, signed on behalf of his tribe a lease giving the Foster Oil Company the exclusive right to prospect for oil on the lands of the Osages. There were dry holes, reorganizations of the company, and assignments of leases before oil was discovered, but in 1905 the Glenn Pool was brought in. In two years it was producing more than a hundred thousand barrels of oil a day. This production was surpassed by the Burbank Pool discovered in 1920.

A census of the tribe had been made in 1906 and each individual entitled to membership was given a "headright." There were 2,229 such headrights. Each headright was therefore entitled to 1/2,229 of the income from gas and oil. In 1924, the year of peak production, this came to $12,000. Every individual who was living when the census was made received this much, but since the population of the tribe had decreased since 1906, some Osages had inherited more than one headright. Some individual incomes were as high as $100,000. In 1957 there were still holders of multiple headrights, but there were some who had inherited only a small fractional interest in one. The total income from oil and gas from 1906 through the fiscal year of 1962 was $437,258,298.15, or income per headright of $196,616. Perhaps a good many white citizens wish that their grandfathers had not crowded the Osages onto the barren Cherokee Outlet but had gone there themselves.

The Osages have had their revenge. Whether or not they have used their money wisely is another question.

The Indian occupied land that the white man wanted. He

resisted by whatever means he could, military and political, but his cause was hopeless. He was overwhelmed by what orators of the nineteenth century termed the "westward course of empire."

The westward course of empire was not much kinder to the University of Texas than it was to the Indian. Texans like to boast of the foresight of their founding fathers in making generous provision for their university. An examination of the records compels some modification of this boast. For nearly a hundred years the custodians of the university lands waged a defensive and not uniformly successful warfare against the settler. That the institution now has a sizable endowment fund is due as much to luck as to foresight.

The Congress of the Republic in 1839 set aside fifty leagues of land (approximately 220,000 acres) for the endowment of two colleges or universities. It was later decided to establish one university instead of two, and the one did not open until 1883, but this was the beginning of the endowment fund. It was also the beginning of a contest between the settlers, who wanted cheap land, and the custodians of the fund, who wanted it to yield the maximum of revenue.

The original grant lay principally in the rich agricultural district of Central Texas. Three leagues, however, were in East Texas, some of it in what proved in the 1930's to be the richest oil field in the United States. This land was surrendered in 1850 by a joint legislative resolution which recited that inasmuch as the field notes had not been filed in the General Land Office and inasmuch as the land had been surveyed in Rusk and Smith counties, the university's title should be extinguished and three leagues selected elsewhere. But if this had not occurred, the university would not have profited by East Texas oil, for the land would have been sold long before 1930.

The act providing for the sale of the original endowment lands was passed in 1856. The land was to be sold in 160-acre lots at auction at a minimum price of three dollars an acre. Twenty years' credit might be allowed at 8 per cent interest. Failure to pay interest and instalment when due would result in forfeiture of title, and the land would be resold to the highest bidder. This looked like a satisfactory compromise between the interests involved. Three dollars was about the market value of the land. The interest rate was adequate but not excessive in those times of scarce capital. The fund seemed to be sufficiently well safeguarded against defaulting purchasers.

But these were only appearances. A later commissioner of the General Land Office was to complain that as the lands "embraced the choicest lands in the republic" and as the country settled up, they were coveted and squatted on by homesteaders. The influence of these settlers secured the passage of several acts providing for the subdivision and sale of the fifty leagues. They were cut up into quarter sections and sold to actual settlers at $.50 an acre on ten years' time with 10 per cent interest. The statutes of the state from that date until the adoption of the Constitution of 1876 will show at each succeeding session, " 'An Act for the relief of purchasers of University lands.' "

"The object and effect of these various laws," the commissioner continued, "was to cancel previous obligations of purchasers, remit interest and allow repurchase at the original price. The result was to finally dispose of these valuable lands at a price far below their actual value and to deprive the University of a large sum due for interest."

In 1854 the legislature passed an act providing for a subsidy of sixteen sections of land for each mile of railroad constructed under the terms of the act. Alternate sections in the railroad surveys were set aside for the benefit of the public schools. An act of 1858 gave the university endowment fund every tenth section of this school land. This act remained in force until 1876 and under it the university would have been entitled to 1,750,000 acres of land. This land too was located in rich agricultural regions. But the Civil War came on and after it Reconstruction, and the provisions of the act were never carried out. Even bonds and cash in the university fund were used for general purposes.

In the constitutional convention of 1875 the victory went largely to the homeseeker. The tenth section provision of the act of 1858 was annulled, and in compensation the university was granted a million acres of land in the western part of the state, land clearly beyond the limits of successful farming. The 1,750,000 acres taken from the university was worth at that time about five dollars an acre. The million acres given to the university was worth about fifty cents an acre, a loss of more than eight million dollars.

There was indignation and agitation for an increase of university land, and in 1883 another million acres were set aside, also in the western part of the state. There was no great generosity in this. The university got the land simply because nobody wanted it. Perhaps that is the reason also why friends of higher education were able to incorporate into the constitution a provision denying the legislature power to grant relief to the purchasers of university lands.

The western land, however, was never sold. The cattle and sheep men, the only ones who could use it, were quite willing to occupy it by lease. Such agitation as there was for sale came largely from local politicians who wanted the land on their tax rolls. They were silenced in 1930 by an act that made university lands taxable for county purposes. For a long time the sole income from this two million acres was derived from grazing leases, which yielded $451 in 1885, $77,388 in 1905, $230,160 in 1925, $375,412 in 1945, and in excess of $600,000 in the 1960's. The average annual rental is about 30 cents an acre. Land leasing at this rate is worth eight to ten dollars an acre, and this is more than the convention of 1875 or the legislature of 1883 thought they were giving the university.

But it is far short of what they actually gave. In 1923 the Big Lake oil field was discovered on university land. Since then several other fields all or partly on university land have been developed. At the end of the fiscal year of 1962, oil and gas had added $385,605,650 to the university endowment fund, excluding the one-third of the income assigned to the Agricultural and Mechanical College. Since the supply of oil is exhaustible, income from oil royalties cannot be spent, but is added to the endowment fund. Until 1956 the constitution restricted the investment of this fund to certain classes of public bonds, upon which the yield has been less than 3 per cent. With part of the fund now invested in commercial stocks and bonds, a higher yield is expected.

The university, therefore, is not nearly so rich as is popularly supposed. It seems richer than it is because it is much richer than anybody fifty years ago had reason to expect. It is more than a quarter of a billion dollars richer because it had to take land that nobody wanted; the Osages became wealthy by being crowded off their land; the federal officials unwittingly rewarded the followers of Crazy Snake; and Dorset in anger gave his daughter the best portion of his estate.

The woman of my acquaintance has paid her back taxes, penalties, and interest. Her banker may be right. Oil may come. But she might be more sanguine if her father had given her the land in anger rather than in love.

7

The Role of Chance

ON SATURDAY, August 27, 1859, Uncle Billy Smith, who had been employed by Colonel Edwin L. Drake to drill a well for oil near Titusville, Pennsylvania, felt the bit drop into a crevice. On Sunday, Uncle Billy plugged the bottom of a length of rain spout, lowered it into the well, and drew it out full of oil. Thus the first well ever drilled specifically for petroleum in the United States was a success. It was a success because of a fortunate accident, for subsequent drilling in the immediate vicinity showed that if Uncle Billy had set up his rig only a few feet away in any direction he would have missed the crevice and the oil. Luck was present at the birth of the oil industry, and the role it has since played has been both factual and legendary.

The importance of luck in the finding of oil has never been denied by reputable geologists, and by none has it been more emphasized than by E. DeGolyer. While working his way through the University of Oklahoma, DeGolyer took a summer job that resulted in his returning to school a millionaire. His advice which resulted in the discovery of a great field in Mexico was based on geological observation, but he always attributed the outcome to luck. "I'm just lucky," he told Norman Cousins.[1] He believed that luck adhered to or inhered in some people more than others. So when he interviewed young geologists for jobs, he examined their credentials, but he also asked if they were lucky. He saw prospecting for oil as a game of gin rummy,[2] in which both skill and luck were needed, but in what proportions he could not say. It was possible to win on luck alone, but never on skill alone. Drilling on a location made by the most refined techniques might result in a dry hole. When you prospected for oil you took a calculated risk, hoping that luck would be with you.

It is not surprising that stories of luck, good and bad, some true, though fantastic, and some false, should be told throughout the oil regions of the United States.

Hal Greer, a lawyer in Beaumont, was a great deer-hunter and cigar-smoker. He always lit his cigars with a flaming piece of wax paper to get the full, even light he wanted. Once he went on a deer hunt in Hardin County, Texas, near Saratoga. He killed a buck, hung him to a tree, and dressed him. He went to a small pond to wash his hands. This done, he lit a cigar and threw the burning paper in the water. The pond caught on fire, telling him that the small bubbles he had noticed were caused by gas. Before he returned to Beaumont he found the owner of the land and bought three acres. When oil was discovered at Saratoga an operator called on Greer and leased the land. Some time later the lease-holder reported that he had drilled according to contract and had found no oil. He wished to abandon the lease. Greer asked him how far he had drilled. He said 1,085 feet. Greer pointed out that the contract called for 1,100, and refused to release him until he had drilled another fifteen feet. This was enough to bring in a good producer and one that brought Greer royalties the rest of his life.[3]

A wildcatter who had missed the Healdton field in Oklahoma by about four hundred yards was wondering what to do next. He had made up his mind to go either to Burkburnett or to Ranger, but he couldn't decide which. He flipped a coin which sent him to Ranger, where he drilled three dry holes on the townsite and lost $150,000. The Burkburnett townsite had one of the highest percentages of hits in the history of oil.[4] If he had drilled there, he would almost surely have brought in producing wells. Another man, uncertain where to set up his rig on a farm he had leased, tossed an empty whiskey bottle in the air and drove the stake where it fell. He brought in a producing well.[5]

Whether or not these stories are literal truth, they express an attitude. It is an attitude that results from the near misses and bare hits that are a part of the tradition of oil. A few hundred feet may separate production from nonproduction. One company drilled a 4,000-foot dry hole on Signal Hill. A half-mile away another company later got oil and brought in a field. Oil was discovered in the Dominguez Hill only 660 feet from a dry hole.[6] If John Galey had driven the stake locating the Lucas well 50 feet south of the hogwallow instead of north of it, he would have missed the oil.[7]

Much has been said about the luck of Dad Joiner, discoverer of the East Texas field. One current story is that he made his first location quite near Mrs. Daisy Bradford's house and that she insisted that he move farther away, so that oil, when it came, would not ruin her washing on the clothesline. Joiner lost his first hole and skidded his rig one location to the east and started another. This hole too had to be abandoned and he again moved his rig one location to the east. His third well came in to open up the great East Texas pool, but if he had lost this hole and had moved one more location east, he would have missed it.[8]

The discovery of the great Yates pool in West Texas hinged upon a promise an official of an oil company had given Yates to drill on his land. The Transcontinental Oil Company had leases in Pecos County which included the ranch of Ira B. Yates. This company made an arrangement with the Mid-Kansas Oil and Gas Company to do the drilling. The plan called for the drilling of three wells, but it was provided that, in case the first and second holes were dry, the company might make a cash payment in lieu of putting down the third well. The first two wells were dry, and Mid-Kansas was ready to make the cash payment and give up. Their geologist, Frank R. Clark, urged them, however, to make one more effort, and Levi Smith, representing the Transcontinental Company, had promised Yates a test. Since Clark had failed to find oil on what he regarded as the most promising geological locations, he picked the spot for the third more or less at random. The well came in on October 28, 1926, and when deepened by stages to 1,150 feet, produced 2,950 barrels of oil in one hour.[9]

Chance operates vertically as well as horizontally. A Texas tool-dresser once associated himself with a group who undertook to poorboy a well on a wildcat lease. Some hundred or so feet short of the sought-for sand their equipment broke down and their money ran out. The tool-dresser still had savings in the bank. He agreed to put them in the well for an additional interest. When he got to town the bank was closed and he couldn't find anybody who would cash his check. When he got back to the well the next day he found that the driller had sold his interest and left. The purchaser completed the well and got a producer. He did not need the tool-dresser's capital.[10]

Near Fulton, Ohio, in 1879, a farmer was living in a log cabin on forty acres of swampland, upon which a driller had a lease. The driller didn't have much money to begin with, and when he had drilled into the Trenton sand, where he had expected to find oil, what he did have was all gone. Convinced that the hole would be dry, he offered to sell the well and lease for $2,500. The representative of a major oil company to whom the offer was made asked for time to think it over. He had not got out of sight of the well when it blew in at 10,000 barrels a day.[11] The Midwest Oil Company's well on Iles Dome in Moffat County, Colorado, had penetrated the sand known to be productive at Moffat Dome, but had got no oil. The officials of the company, meeting in Denver, decided to abandon the test, but before the meeting adjourned a telegram arrived informing them that the well they had just voted to abandon was producing oil.[12]

George Hooks went to Ward County, Texas, leased some land, rented a drilling rig, and hired a crew. He was operating on a shoe-string and was having a hard time meeting his payroll. When he was down about a thousand feet, he had a chance to sell out for enough to pay his debts, and he decided this was the best thing to do. The purchasers deepened the well fifteen feet and brought in the Bennett field. After this experience, Hooks's advice was, never abandon a well until you have gone fifteen feet deeper.[13]

His and similar experiences would indicate that it is to the advantage of the wildcatter to have plenty of money. But even this rule has had its exceptions. The first well that O. W. Killam and associates drilled in South Texas developed a light showing of oil at 1,400 feet, but there seemed to be no chance of making it a commercial well. They moved about a quarter of a mile east, got a better showing, and decided to case the well. They set the casing on a shoulder, made by reducing the size of the hole, and did not cement it. When they tried to test the well the casing gave way and dropped sixty feet below. They decided that there would be less risk in drilling a new well than in trying to salvage the old one. They moved the rig sixty feet and began drilling again. By this time money was running low. Coal for the boiler had to be hauled eighty miles. They economized by mixing mesquite wood with it. When they were down 1,413 feet, they had about a wheelbarrow full of coal left. The driller said that that would be enough to set the casing, but that if he drilled deeper he would have to have more fuel. Killam told him to set the casing. This time they cemented it. When they drilled out the plug the well made 30,000 barrels of oil in a day. Killam said that if they had had plenty of fuel, they would probably have gone through the oil sand without knowing it and the Mirando field would not have been discovered. Certainly it would not have been discovered by Killam. He could not have raised money for a fourth well.[14]

Two motifs stand out above all others in the lore of luck in finding oil. One of these might be called the lucky accident. In 1918 the Humble Oil and Refining Company had completed the drilling of Gulf Fee No. 4 at Goose Creek. The tool was being lowered to perforate the casing at 3,000 feet, the depth of the known producing sand. The tool, however, stuck between 2,200 and 2,300 feet, and the casing was perforated about 700 feet higher than was planned. The well came in at 10,000 barrels, revealing a producing sand that had been passed in drilling.[15]

The most common form of the lucky accident is that in which the well for some reason, often a breakdown in transportation, is drilled somewhere other than the planned location. It is a producer, but a well later drilled on the original location is dry.

D. L. Wolfe was a rancher in Archer County, Texas. He was also something of an amateur oil witch or doodlebug man. When an operator offered to lease his land, he let him have the larger part of it but retained a part for himself. On the part that he had retained he made a location—not that he had any plans for immediate drilling. The stake merely marked the spot where his wiggle stick told him to drill if he ever drilled at all. When the operator sent in his rig, the crew could not find anybody to point them to the location. They found Wolfe's stake, and assuming that it marked the place, they set up the rig. They had drilled over a thousand feet before their mistake was discovered. The oil man went to Wolfe and told him what had happened, and Wolfe agreed to exchange the land for a part of the lease. The well was a producer. The original location was later drilled, and the result was a dry hole.[16]

More often the mislocation is the result of a breakdown in transportation. This story, observed Samuel W. Tait, Jr., in his *The Wildcatters*, "has been related about every oil field I know, whether in sandy desert, boggy swamp, muddy prairie, or rocky mountains, and it has probably happened in every one of them." The one instance that Tait vouches for happened in the sandy desert of California in 1895. C. A. Canfield and J. A. Chanslor of Los Angeles took a lease in the Coalinga district. As they were hauling their drilling equipment through the deep sands, a wagon broke down several miles from their location. Instead of unloading the wagon and repairing it, they decided to set up the rig and drill. The result was a three-hundred-barrel well which led to the development of the Coalinga district.[17]

There is the case of the fraudulent promoter who turned honest in spite of himself. He was employed as an office worker by a company that was more interested in selling stock than in finding oil. As checks and money orders came in, it occurred to him that his income need not be limited to his modest salary. He secretly copied a thousand names from the company's sucker list and sent out literature which he had got up in the style of his employers. The response was greater than he had anticipated. He sent out another thousand letters, and again the money came rolling in. Then he began to worry about postal authorities. He had claimed to have a lease and had promised that drilling would start soon. If he didn't drill he might end up at Leavenworth. He could put down a well and still have money to the good. He went a long way from production where leases were cheap, and hired a man who called himself a geologist to make a location. By the terms of his lease he must start drilling in ninety days. He had difficulty finding a rig and more difficulty getting it to the lease, for the rainy season had begun and roads were flooded. His time was about up and there was a swollen creek between the rig and the location his geologist had staked. But he was on the corner of the lease, so he had the rig set up. He brought in several paying wells, but the one that was later put down on the original location turned out to be dry.[18]

There are several occurrences of the lucky breakdown story in Texas. Of these I have been able to verify only one. The big Cooke field in Shackelford County was brought in in 1926, and has produced oil worth over twenty-five million dollars, and is still producing several hundred barrels a day. It would not have been discovered at the time if the first well had been drilled on the location made by the geologist. He had driven the stake on the top of a hill, which proved to be too steep for the trucks that brought the equipment. Instead of building a road, the operators decided to drill at the foot of the hill. The location made by the geologist was afterward drilled and the hole was dry.[19]

This experience is little known, but nearly everybody in Oklahoma and Texas knows the legend of the Fowler Farm Oil Company and those in other states have had opportunity to read about it in books, magazines, and newspapers, including the *New York Times.*

S. L. Fowler, owner of a large and rich cotton farm in the Red River Valley just north of the little town of Burkburnett, Texas, in 1918 decided to give up farming and become a ranchman. He figured that the proceeds from the sale of his farm would buy sufficient grassland to sustain a herd large enough to make him a good living. The only difficulty was that Texas has a community property law, and Mrs. Fowler would not consent to the sale. She was quoted by one journalist as saying:

I believe there's oil under our land, and I won't agree to dispose of the farm until a test is drilled. If oil isn't found, I'll sign the papers; but the people of this section have undergone so many hardships with such patience that I just know there is something good in store for them.[20]

Finding his spouse adamant, Fowler decided that the only way to sell the farm was to meet her condition, so he called on his neighbors for help. He figured it would take $12,000 to put down a well deep enough to satisfy his wife. He put up a thousand dollars and his friends subscribed varying amounts until the sum was raised. One of the subscribers was Walter Cline, a drilling contractor who happened to have an idle rig in North Texas at the moment. In exchange for a thousand-dollar interest in the venture, he agreed to furnish the rig and services of his drilling superintendent.

According to legend, a geologist was employed who staked out a location in the cotton field. The wagon hauling the first load of equipment bogged down in the sand. "Oh well, unload her here," Fowler said, since for his purpose one place was as good as another.[21]

Other versions lay the blame on the teamster. A *Saturday Evening Post* article says he somehow missed the stake and unloaded the rig at another place. Instead of having him unload and move it, the drilling contractor, impatient to get started, set the rig up, saying, "On a big farm like this one location is as good as another." A well drilled on the original site was dry. Had not the teamster made the mistake, "the whole Burkburnett development would have been delayed, and possibly not started at all."[22]

A *New York Times* story attributes the Fowler discovery to "brute accident."

Fortunately, the teamster dumped the heavy rig at the most convenient place on the farm. Rather than go to the trouble of reloading and trucking it to the place he had chosen, the driller decided to go ahead here. He struck oil—and every $100 of the farmer's pool repaid its investor 120 times over. Later a well was put down at the place originally picked by the driller—and it proved to be a dry hole.

Out of that careless teamster's mistake came the great Burkburnett field—which made several cattlemen into oil barons and strewed the plains with boisterous towns.[23]

The well came in and opened the Burkburnett field. After a few months the Fowler Farm Oil Company sold out for $18,000,-000, paying the shareholders $15,000 for each one hundred invested.

The figures are history and a part of the legend is true. Neighborliness did enter into the formation of the company. Sand had something to do with the location of the well, and a well drilled later north of the discovery produced only salt water. The rest is imagination.

As Walter Cline remembered, the project began to take shape one day in front of Luke Staley's drugstore, "where we usually congregated and did our whittling and settling really heavy problems."

S. L. [Fowler] broke the news to the bunch [Mr. Cline continues] that he wanted to dispose of his cotton land and get him a piece of ranch land but that his wife thought they ought to drill a well on the land and that he didn't have money enough and he was going to have to get some help, and he wanted to know if any of us would be interested. Well, there was nothing particularly favorable to encourage anyone to want to spend any money or time or effort on the Fowler land. On the other hand, there was nothing that definitely condemned it. So we sat around and decided that just as friends and neighbors, we'd just do our boy-scout good turn by putting in a little, not enough to hurt any of us, but maybe enough to poorboy a well down. . . .

And we got enough money committed to look like we could afford to drill a well. The question of location then came up and we decided there wasn't any use in pulling a whole lot of sand, going way out in this cotton patch, and we'd drill it reasonably close to town. So we went right north of the hog pen that was east of S. L.'s house with a lane through there and a gate, drove out in the field where we had driven a stake and drilled there.

I'd like to interject here that statement that there've been very few discovery wells brought in that the myth hasn't started that they decided to drill a well on a given farm or on a given part of ground and they started out and it rained like the devil or they broke the wagon wheel or the truck broke down and they were already on the property where they wanted to drill and they were a half-mile or a mile from the location, and they said, "Oh well, hell, let's just unload it. One place is as good as another. We're on the right land. We'll drill it here."

Well, now I heard that. I've heard it about the Fowler well and I've heard it about practically every discovery well that's been drilled in Texas in my time, and I have yet to find a single instance of where that's true. I just don't think there's a bit of truth in it. I know it's not true so far as the Fowler well is concerned. We drilled the Fowler well right where we intended to drill it, and right where we drove our stake. And that definitely settles the Fowler well.[24]

Another and more elaborate version of the lucky breakdown story is the one concerning Santa Rita, the discovery well in the Big Lake field, the first of the oil fields to be opened on the endowment lands of the University of Texas.

The earliest published version of this story that I have found is that of Owen P. White in the *New York Times* for May 3, 1925.

When the lease was signed Krupp [Hymon Krupp, organizer and president of the company] went to work. He hired expensive oil experts ... to go out and locate the proper anticlines for him and then after this had been done, busied himself in getting together enough money to drill the well.... The months flew by until the date on which the lease would automatically expire was dangerously near. Krupp redoubled his efforts and finally, with only a few days to go, he took the road with three trucks loaded with drilling equipment.

When Krupp was still several miles away from the precious stake, which had cost him several thousand dollars to have driven in the ground, one of his trucks broke down completely, and—there he was! At that time, as the story goes, he had only two days left. What should he do? There was no possibility of being able to reach the desired destination with his outfit. The breakdown had occurred on land covered by his lease and so, with no high-salaried geologist at hand to advise him, but merely because the ox was in the ditch and he had to act at once to prevent forfeiture, Krupp set up his rig at the scene of the disaster and went to work.

Another version of the tale was included in the general report of the Sun Oil Company covering operations in West Texas and New Mexico up to November 1, 1929. A paragraph reads:

An unusual incident occurred when the Big Lake Field in Reagan County was discovered in 1923. Location for their first test was made some two miles westward, but owing to a breakdown while transporting the materials to the location, they unloaded just where the incident happened, and this well, while it led to a small producer, led to the discovery of the Big Lake Field. Had the first test been drilled where the location was originally made, possibly new chapters would have been written regarding the field. Subsequent tests have proven their first location would have been a failure.[25]

A writer for the *Daily Texan*, student newspaper of the University of Texas, for February 9, 1940, told the story as follows:

The site, in the southwestern part of Reagan County, at which the first oil well was completed in May, 1923, was chosen purely by accident. The drilling party, headed by Frank Pickrell, was bogged down. Since the lease was to expire in a few hours, members decided to drill where they were stranded.

They drilled not on their own lease, but on a part of the 2,000,000 acres provided by the Constitution for the University endowment fund and now known as the Big Lake Field.

But for a real professional handling of the story, we turn to the *Austin Statesman* for January 23, 1940.

The ragged country with its old worn jutting hills, crouched beneath the terrible drenching from the rains. Little rivers ran where dry gulches with their platted grasses formerly cut through the terrain.

In slicker and chewing the end of an old cigar, a man named Frank Pickrell peered from the switch house into a torrent of rain, walked impatiently back and forth.

"All right, boys," he said, "I've got a lot of money tied up in this. We've got to take a chance. We've got to get this machinery going."

And the boys got up and in the heavy rain began loading the rig and drilling machinery on the cumbersome wagons. In a little while they got started.

Mud clawed at the wheels and sucked at the mules' feet. The animals grunted and strained at the traces and the wagons creaked through the slime over the treacherous roads.

"Just seven miles to go," encouraged Mr. Pickrell. The rain poured. The men cursed and cracked whips, and wiped the mud from their eyes.

The geologist had said that oil could be found on a certain spot. Mr. Pickrell was determined to reach that spot. Then the rain came down in torrents. It almost hid one team from another. West Texas had never seen it rain like that before.

Up front there was much cursing. Hazy figures floundered here and there. Mr. Pickrell stalked up front. The lead wagon was mired. There wasn't any use, the straining mules could not budge it.

"We'll have to wait," said Mr. Pickrell. The boys huddled together to wait. The skies were puffed and swollen with clouds, and the rain chattered along the gullies and around the wagons.

They waited all day. Mr. Pickrell knew his West Texas. "Boys," he said, "this thing'll keep up. Another day and we'll be here two weeks getting to that place. Unload her here. We'll dig our well right here."

Now the testimony gathered by Schwettman from Pickrell and other participants in the event is to the effect that (1) it was not raining when the equipment was being moved; on the contrary, the ranchers were complaining about a long dry spell, (2) horses rather than mules or trucks were used to draw the equipment, (3) there was no breakdown, (4) the well was located where the geologist had driven the stake.[26]

Pickrell had previously certified that the well had been located at the stake driven by the geologist and that it was upon the recommendation of the geologist that the well had been drilled. Moreover, the breakdown story became an issue in a lawsuit in 1926. The court found the story false and so stated in its judgment. This, however, did not dispose of the legend. Three of the four written versions quoted came after the decision, and Texans still talk about the lucky breakdown that brought oil to their university.

Almost as widely diffused as the story of the lucky accident is that of the disobedient oil driller. There are numerous examples of oil's being discovered through the failure of the men in the field to carry out the order to cease drilling. Working on a well owned jointly by the Lakeview and Union Oil Companies in California in 1910 was a foreman, "Dusty" Woods, whose nickname reflected his drilling record. It looked as though that record would be maintained until March 15, 1910, when he reached the oil sand, and the officials ordered him to put the well into production. Instead, he drilled 47 feet deeper and brought in an 80,000-barrel well.[27] The Royalite Oil Company was sinking a test in Turner Valley, Alberta. After a showing of oil had been encountered and the drill had moved down to 3,800 feet into a thick limestone, orders came from headquarters to abandon the test. The drilling superintendent ignored the order. A local official of the company, by telegram, informed headquarters in Toronto of this insubordination, but before the superintendent could be fired the well blew in, making seventeen million feet of gas a day, from which six hundred barrels of condensate could be extracted.[28]

One form of the story of the disobedient driller is the legend of the million-dollar drink. This is no Coal Oil Johnny episode in which a million dollars is spent; instead, a million dollars is gained, though not for the celebrant. The big brass in a distant city decide to abandon the well. But the driller or someone in the chain of authority between him and the president fails because of drunkenness to see the order carried out. Drilling goes on a few hours longer and a million-dollar sand is tapped.

Whether or not this ever happened I do not know. It did not happen in two reputed instances in Texas. One of these concerns the McClesky discovery well at Ranger. Neither Frank Champion, the driller, nor W. K. Gordon, the resident engineer of the Texas and Pacific Coal Company, was drunk. What happened was that the officials of the company had telegraphed Gordon advising him to quit. Gordon wired back for permission to drill further and was authorized to use his own judgment. By October 21, 1917, Gordon had either given up hope or deferred to his superiors, for on that day he instructed the drillers to work until night and shut down. About two o'clock in the afternoon the well blew in.[29]

The classical Texas example of the million-dollar drink, like that of the lucky breakdown, comes from the Big Lake field and involves Frank Pickrell. In 1928 a deep test known as University 1-B was being put down with cable tools. At a depth of 8,245 feet the bailer hung, the sand line parted, a long and expensive fishing job ensued, and the company decided to abandon the well. The reason the order was not carried out, according to a widely circulated report that reached print in the *San Angelo Standard-Times* for May 28, 1933, was the million-dollar drink.

Late in Nov., 1928 Frank Pickrell, Texon vice-president, ordered a halt at 8,343 feet. The deep wildcat had cost over $100,000. Cromwell reported that the formation looked promising and argued successfully that drilling continue. On Saturday, Dec. 1, the bit reached 8,518 in black lime. Pickrell again telephoned and this time insisted that work stop. Cromwell stopped to revive his spirits with what someone later aptly termed "a million dollar drink," and decided he could notify the crew the next morning. Meanwhile the drill kept pounding.

Early the next morning the driller phoned Cromwell that the well was spraying oil. "Hit 'er another foot," Cromwell instructed and No. 1-B began to flow for a new world's record depth of 8,525 feet.[30]

Schwettman also investigated this legend. What basis it has in fact is indicated by a letter received from Waldo Williams, chief driller. Williams said that the "big boys had a meeting in New York and decided that $140,000 was enough to spend on a nonpaying well." They ordered Pickrell to stop drilling at 8,500 feet. Pickrell then called Cromwell by telephone and gave the order. Cromwell then talked over the prospects of the well with Williams, and was so confident of success that he said instead of giving the order he would disappear for a few days. On the second day of December the well flowed with forty barrels of oil.

Williams began trying to locate Cromwell. After two days he was found at an editors' convention in Sweetwater. A few hours after he received the message, he was on the location directing the completion of the well.

Legends are curved mirrors reflecting but distorting the truth. If luck played no part in the oil business, stories like that of the drunken driller would not be told. And yet luck is of less importance than is generally supposed. Looked at in its entirety, the history of the oil industry is the history of a development that seems all but inevitable. By the middle of the nineteenth century the chief source of illuminating oil was rapidly waning. The whale had about disappeared from the Atlantic and our whaling fleets were making the long voyage around the Horn and into the Arctic regions in search of a dwindling supply of blubber. In 1850 James Young of Scotland produced a satisfactory lamp oil by the distillation of coal. By this time, too, the chemical properties of petroleum were becoming known. Its main use heretofore had been medicinal. One of the men who sold petroleum in small bottles as a cure for practically every disease was a Pittsburgh druggist named Kier, who secured his oil from a well that had been drilled on his father's place for salt. Kier succeeded in distilling petroleum and securing a product comparable to the coal oil of James Young. Hitherto petroleum had been obtained only from oil seeps and springs and from wells where it had accidentally been found, but now that it was known to have commercial possibilities, it was inevitable that somebody would begin seeking it by drilling. If Drake had missed the crevice the completion of the first oil well might have been delayed for a few months or a year, but not for much longer. And viewed in perspective, the enormous expansion of the oil industry that followed Drake's discovery seems inherent in the logic of events. By the time electricity became a serious competitor of kerosene, the internal combustion engine had ushered in the motor age.

If Dad Joiner had skidded his rig over one more time and missed the East Texas field, it would have been discovered one month later, for only a few miles to the north another well was already drilling, and if Gordon had shut down the McClesky well when his superiors suggested it, the Walker well in the same field, which at first produced only gas, but later blew in as an oil well, would have been the discovery well of the Ranger field. These events made a lot of difference to Joiner and Gordon and McClesky, but little to the national economy.

Blindfold a hundred men, set up a target within range, and indicate its direction; give each man a rifle and tell him to shoot once at the target. It is almost certain that some one or two will hit it, but nobody can predict which one or two it will be. Give one man all the shots and he is almost sure to hit it, but it is impossible to predict whether it will be with his first shot or his hundredth. Let us assume that the contest has been conducted like an old-fashioned shooting match, each man paying a fee to enter and the best shot getting the reward. Then we have something somewhat analogous to the oil industry. What a man's chances of finding oil by mere random drilling are, I do not know, for while men have sought oil on mistaken assumptions, they have never drilled completely at random. Perhaps the nearest approach to it has been drilling on the advice of doodlebug men, one of whom had oil discovered on the sixty-sixth location he made. Geology and related sciences have not entirely removed the blindfold, but they have made it somewhat transparent, and the remaining risk diminishes as the individual or company acquires resources to buy more shots. Thus the wildcatter who had missed the Healdton field by a few hundred yards, and who flipped the coin that brought him to Ranger where he drilled three dry holes, went to Desdemona where his persistence was rewarded with a gusher.

Besides luck variable factors are the development of new techniques and the use of these techniques to the point of diminishing returns. Records indicate that over the long haul a stable company acting upon the advice of a competent geological staff averages slightly better than one hit out of four shots for all wells drilled, the majority being on structures already proved; and one out of nine or ten for rank wildcats. When all wildcats drilled for whatever reasons are taken into account, the chances are reduced to one in fourteen.[31]

II

Popular
Stereotypes

THE FOLK WAY *of reducing the complexities of human character to manageable concepts is simplification. There is first the
unconscious assumption that all or most members of a group, an
occupational group for example, exhibit a few common traits.
These traits are extracted and thus a stereotype or public image
is created. The accuracy of this image is not to be accepted
without question, not only because it is an oversimplification,
but also because it is based upon the most conspicuous behavior
of the most conspicuous, though not necessarily the most representative, individuals. The cowboy who got drunk and shot up
Dodge City became the norm, not the one who behaved with a
modicum of decorum.*

*New industries generate new occupations, and new occupations
generate new stereotypes. But the new stereotype will have recog-
nizable antecedents. The geologist is related to the academician,
the oil promoter to the Yankee peddler; the shooter and the
driller bring to mind images of the keelboatman, the trapper,
the miner, and the cowboy; the landowner is the suspicious countryman of tradition.*

8

The Geologist

THE STORY of the lucky breakdown suggests that the status of the geologist in the first thirty years of this century was somewhat below that of hero. In the rejoicing of a community when oil came, he was likely to be overshadowed if not forgotten. The oil field that he mapped was not named for him.

Samuel H. Tait complains of a "convention of anonymity about all oil discoveries which is observed by all writers," who, he says, fail to disclose that back of every big discovery "was a single man, usually a geologist, with the courage to fight for his convictions at the risk of losing his professional standing."[1] There is no reason, however, for believing that there has been a conspiracy on the part of reporters and popular historians to belittle the role of geology in the development of the oil industry. They brought to their writing certain presuppositions about the qualities which gave interest and significance to the events reported. That is, they assumed that they knew what their readers wanted: conflict, humor, sentiment, action, and the like. These were the qualities they failed to find in the work of the geologist. His conflicts were conflicts of ideas. His battles were waged within a small group, not in the open forum. If they came to the attention of the reporter at all, he did not know how to make a feature story of them. The result of his writing was more to confirm the popular ideas of geology than to impart accurate ones.

The attitude of the farmer or rancher toward the geologist was ambivalent. If he were known to be looking for oil, he was welcome. If questioned about the prospects, he was evasive, but if you watched him work carefully, you would see that he kept coming back to a certain place, and that would be where it looked best for oil. If his report was unfavorable and nobody offered to lease your land, he probably didn't know very much about his business. If his report was favorable and somebody leased your land and put down a well and didn't get oil, the geologist was unquestionably right, but the driller was incompetent or dishonest. Maybe he had passed the oil sand without knowing; maybe he didn't drill to the specified depth, made a false report, and collected his money; or more likely he was bought off by a rival company, probably Standard Oil, and plugged the well.

Nobody knows when the term "rock hound" was first applied to the geologist, but it might well have come from the country folk. M. G. Cheney noted that the people who watched him work thought that he roamed over an area in the manner of a coon dog looking for a scent, and when they saw him return to a point of reference, they thought of a coon dog returning to the place where he had lost the trail.[2]

And Charles Gould reports the following story:

...an old farmer in southern Kansas was surprised and concerned to see two men with surveying instruments out on a rocky hill in his pasture. He asked a neighbor if he knew who the intruders were.

"Oh, them's rock hounds," said the neighbor. "They've been chasing that ledge of limestone clear across the country. Yesterday Bill Jones, he started to chase 'em off'n his place, but they told him they was huntin' for oil, so he let 'em stay. Says he thought if they could locate oil on his farm, he'd better let 'em do it."[3]

If "rock hound" originated among country folk, it had for them no derogatory implications. A hound is a smart animal.

Among oilmen it probably did. Operators were slow to recognize the validity of geological science, or at least of its usefulness to them. Their saying, "Geology never filled an oil tank," persisted long after it had ceased to be true.[4] John Galey in his early Texas years said that the only reliable geologist was Dr.

Drill.[5] Another successful wildcatter used to say, "To hell with geology. Let's go dig an oil well!"[6]* Benjamin Coyle though a geologist's favorable report was like a shot of dope. "It just spurts him [the oilman] to go ahead ... you don't like to go out and start anything if somebody don't give you a shot in the arm." This was the chief value of geology. He had some mighty nice friends who were geologists and he used to tell them that if he ever found a guy that knew as much as they did, he was going to marry him. At least, "he'll think I'm married to him, I'll stick so close to him."[7]

One way to belittle geology was to give fantastic accounts of how decisions on drilling locations were made. One old-time oil-man told Charles Gould that "his favorite way of locating oil was to tie a tin can to a dog's tail and start the dog on a run across the prairies. Where the can came off, there he would drill."[8] And Lew Allen reported in 1922 that "although admitting that most of the recent strikes have been located by geologists, many oldtime operators effect to scorn the 'rock hounds' and use their own original methods." He quotes one man as saying,

Not that I don't believe in geologists. I always use one. I'm that big a fool. Pay him $50.00 a day to chip rocks and write reports. But when I get ready to start, I take a Negro and blindfold him, turn him around three times, and let him throw a silver dollar as far as he can. Where the dollar falls, if I can find it, is the spot where I drill.[9]

One operator was told by his geologist that he had located the well off the structure. His reply was, "Well, skid the structure over a little."[10]

The drillers and tool dressers and roughnecks were as con-

---

*This attitude, however, was not universal. The Rio Bravo Oil Company, a subsidiary of the Southern Pacific Railroad, was the first oil company to set up a geological department. The department was organized by E. T. Dumble in 1897. (See Edgar W. Owens, "Remarks on the History of American Petroleum Geology," *Journal of the Washington Academy of Sciences*, XLIX [July, 1959], 256-60.)

temptuous as their employers. "The average fellow in the oil field," reports one geologist, would say, "There goes one of those scientists," or, "Here comes another one of those scientist so and soes."[11] Ex-driller Billy Bryant spoke for many of his trade when he said that if he were a big oilman he would hire a geologist, one of the best he could find. "And every time I got me a section of land, I'd let him go out and make a location, and if he made the location on the south corner, I'd go to the north corner and make a well."[12]

Such appraisals of the geologist may be accounted for in part by the fact that the early discoveries were made without his help. The pioneer seekers of petroleum had little need for geology. They looked for oil springs or oil or gas seepages, clearly visible on the surface. "Nearly every producing region (*petroleum province)*" says A. T. Levorsen, "was discovered as a result of drilling prompted by the recognition of nearby surface or subsurface showings of gas, oil, or asphalt."[13] Among the subsurface indications the most important was the encountering of oil or gas in wells drilled for water.

Two epoch-making discoveries will come to mind immediately. Drake drilled near an oil spring. Once the province was discovered, geology was able to trace it into West Virginia. Patillo Higgins' interest in Spindletop began when he noticed gas seeps. The well proved to be on a salt dome, and led to a search for other salt domes, but in this search science was not especially helpful until the introduction of geophysical instruments. Before that time, a layman could do about as well as a scientist.

Equally important in accounting for the tardy recognition of geology was the practical man's contempt for the theorizer, the doer's contempt of the thinker. Even before Drake's discovery in 1859 a Canadian geologist had formulated a structural theory to account for a series of oil seepages, a theory which by the end of the 1860's had been elaborated and accepted by many geologists.

But this early work was done by academic geologists more interested in advancing their science than in finding oil.[14] It was not until 1883 that the first consulting petroleum geologist, I. C. White, opened his office. Writing in 1892, he expressed gratification "in having assisted in removing this stigma [expressed in the saying "Geology never filled an oil tank"] from our profession."[15] But the stigma was not to be completely removed for a long time. As late as 1952 one successful independent operator felt that a geologist's report should be studied with great caution. The geologist is a scientist, accustomed to thinking in scientific terms, and if he finds "even a suspicion of a structure . . . he, perhaps unconsciously, takes the position that a well ought to be drilled to find out what's there."[16] In brief, there is an inherent conflict of interest. The operator wants oil; the geologist wants information.

Yet in the widely publicized classical examples of geological error, the advice has been *not* to drill. Any geologist with long experience in petroleum will confess to errors in the interpretation of data, and he knows the uncertainty of finding oil even on the most favorable structure. Charles Gould, who has been called the father of petroleum geology in the Southwest, and who certainly found his share of oil, used to say that any anticline was worth drilling,[17] but he knew that not every anticline contained oil. The salt domes of the Gulf Coast exhibited no surface outcropping by which they could be defined, nor did stratigraphic traps such as the ones underlying the great East Texas field. In the Mexia region Colonel Humphreys disposed of a part of his holdings upon the recommendation of a geologist, who said it was not on the structure. When the company that had acquired the leases brought in a well, Humphreys wired the geologist, who was in New York, to hurry back to Texas. They had skidded the structure over.[18] Wallace Pratt, one of the great geologists, who had much to do with the success of the Humble Company, had had the same opinion, but he solved the geological problem in time for his company to make profitable leases. He seems to have miscalculated the angle of the fault, and thus oil was produced on what on the surface appeared to be the downhill side.[19]

The successes of geology were taken for granted. It was the failures that passed into tradition.

They have received embodiment in a widely diffused narrative. A local citizen, hoping to strengthen the economy of the community, or a wildcatter, hoping to make money, becomes convinced that there is oil under a certain terrain. The reasons for this conviction vary all the way from religious faith to a new and unorthodox geological theory. A geological opinion is obtained, sometimes from a "government" or "state" geologist, sometimes from a consulting geologist, sometimes from the staff of a big oil company. The report is emphatically unfavorable, and often the geologist offers to drink all the oil found.

But bolstered by his own theory or by his mystic faith, and supported by the local citizens, the wildcatter moves in his equipment and proceeds to drill. The condemnation of the location by geology, however, makes it exceedingly difficult to find the risk capital necessary to finance the drilling. The first attempt to drill a well is usually a failure, either because the equipment is inferior or because the producing sand has been missed by a few feet or a few hundred feet. Additional money has to be raised, but the local people have lost faith and refuse to make further investments. Every legitimate method of financing is resorted to. Finally, when money and credit are down to the last dollar, the well comes in. A new field or perhaps a whole new province has been discovered. The big companies that have refused to help finance the well rush in and lease land at rentals a hundred or a thousand times those prevailing before the discovery. The geologist, instead of drinking oil as promised, says, "Ah!"

This is the pattern, the archetype, in which the popular imagination, aided by popular journalism, tends to fashion the event, though not every element is present in every version. It is not entirely without truth, but upon investigation of particular versions the truth proves to be somewhat more complex and somewhat less disparaging to geology.

Although oil and gas seeps had been known to exist in Texas since 1543, in 1900 there was only one oil field, Corsicana, operating in the state, and its production in that year was only 829,544 barrels. On January 10, 1901, the Lucas Gusher came in, and by the end of the year Spindletop field had produced 3,593,113 barrels. The next year its production was 17,420,049 barrels, and Texas took rank among the leading oil-producing states.[20]

The story of this field need not be told in detail.[21] We are here concerned with the role of geology in the first discovery of commercial production on a salt dome. C. A. Warner cites 1892 as the date and Spindletop as the place of the first attempt in Texas to utilize geology in the search for oil,[22] implying an acceptance of Patillo Higgins' lifelong claim that he had geological reasons, other than surface gas, for believing that oil could be found. Higgins had studied a publication of the United States Geological Survey, and had perhaps identified the low mound four miles south of Beaumont as an anticline; but unless it is among his unpublished papers, he has left no detailed account of his theory. He did believe that gas in contact with sand and shale would cause them to turn to rock, and thus form a roof for an oil trap.[23] But if the memory of one of his fellow-townsmen may be trusted, he also witched the hill with a peach limb.[24]

Captain Lucas, who took over when three attempts to penetrate the quicksand had failed and Higgins was at the end of his resources, may also be said to have been prompted by geology. As a mining engineer employed by a salt company in Louisiana, he had encountered sulfur and showings of oil in association with salt. He correctly surmised that the low mounds on the coastal plains might be subterranean plugs of crystallized salt which might form oil traps.

Aside from Higgins and Lucas, four professional geologists passed judgment on the hill as a prospect for oil. Two judgments were favorable and two were unfavorable.

After the failure of the drilling contractor to attain the specified depth, Savage Brothers of West Virginia offered to put down a well for a 10 per cent royalty. They were advised by a "practical" geologist named Otley. Otley belonged to the trend, or vein, as opposed to the structural, school of geology. He held that deposits were in veins running from mountain ranges. He predicated a vein running from the Rocky Mountains to the Gulf of Mexico, thus providing an abundance of oil on the Coastal Plain. But in spite of Otley's conviction, Savage Brothers gave up their holdings following the failure of their second attempt to drill through the quicksand.[25]

Higgins wrote to Robert Dumble, chief state geologist of Texas, asking him to come to Beaumont and investigate the mound. Dumble could not come, but sent his chief assistant, William Kennedy. Kennedy evidently took the assignment seriously. He could find no reason for believing that oil would be found. He said that rock impervious to oil must overlie an oil deposit, and that he had found that a well in Beaumont, only four miles away, had been drilled to a depth of 1,400 feet without encountering such rock. He thought Higgins was wasting his money, and to protect others who might be inclined to invest, he published an article in the Beaumont paper.[26]

When Captain Lucas had taken over and exhausted his resources, C. Willard Hayes, of the United States Geological Survey, appeared in Beaumont, apparently upon his own initiative. He said that there was no precedent of oil's being found in the unconsolidated sands, shales, and gravels characteristic of the Coastal Plain. And like Kennedy, he added subsurface evidence. In search of artesian water, the city of Galveston, some sixty miles from Beaumont, had drilled a well 3,070 feet, and had encountered nothing to indicate the presence of oil.[27]

The fourth geologist to express a judgment on Spindletop was Dr. William Battle Phillips, professor of field geology at the University of Texas and director of the Texas State Mineral Survey. He came to Beaumont, inspected the area, and talked to Lucas. He decided that the chances for oil were good. He suggested that Lucas approach Guffey and Galey, an oil firm of Pittsburgh which had holdings in Corsicana. To make the approach easier he gave Lucas a letter of introduction to John Galey.[28] Galey at that time had little faith in geology, and it is doubtful whether Phillips' letter was decisive. Yet it was a geologist who brought Lucas and Galey together and led to Guffey and Galey's financing the discovery well.

But for many years Phillips was forgotten and Kennedy and Hayes were remembered, and geology was said to have failed initially in Texas; and for that reason it was not employed in localities where it could have been useful.[29]

The Commonwealth of Massachusetts had filed a suit against Edgar B. Davis of Luling, Texas. Davis had lived in Luling since 1922, but the commonwealth maintained that he had been a citizen of Massachusetts until 1926 and that he owed the state income taxes on a sum estimated as high as twelve million dollars. When a lawyer called on one citizen of Luling in search of evidence that Davis was not a legal resident of Texas, he was promptly informed that any man who said Davis was not a bona fide citizen of Luling, Texas, was a lying son of a bitch.[30]

This was hardly legal evidence, and Massachusetts eventually obtained a judgment, but it is indicative of what the people of Luling and the surrounding country thought of Davis. He was emphatically a hero. Not only had he brought wealth to the community by discovering oil, but he had distributed his personal fortune in giving generous bonuses to his employees, in sponsoring art, in building community clubhouses, and in chartering and endowing the Luling Foundation for the betterment of agriculture.

And in the search for oil he had spent considerably more than a million dollars.

The people of Luling in recounting his exploits to Stanley Walker[31] repeated the legend of his single-handed struggle to find oil. They reported that "most geologists and most of the oil companies were convinced that there was no oil in Caldwell county." Of the same import had been an article in a historical edition of the Lockhart *Post-Register* for August 3, 1936, in which the only reference to geology was the statement that Davis had begun his seventh and first successful well against the advice of his geologist. There was no reference to the origin of the search for oil.

This search was quite typical of the times in that it was initiated by local men motivated at least in part by a desire to add a new resource to the not overly prosperous agricultural economy. Two lawyers, Norman Dodge and Carl C. Wade, asked a geologist, Verne Woolsey, to look for a place to drill. Woolsey remarked that they were sitting on a fault at the moment. He located the fault plane, and Dodge and Wade took leases. Wade then went East seeking capital to drill.[32] He succeeded in interesting Oscar Davis, who took stock amounting to $75,000 in the newly organized Texas Southern Oil and Lease Syndicate, and asked his brother Edgar to manage his interests for a third of the profits. When the capital of the syndicate was exhausted, Edgar Davis paid off the stockholders, including his brother, and organized the North and South Oil Company to continue exploration.

He had been so confident of success that he had begun three wells. After drilling a total of six dry holes and spending the fortune of a million and a half dollars he had made in rubber plantations, he made, against the advice of his geologist, a location of the Raphael Rios property. The well came in August 8, 1922, and led to further development in Caldwell and Hays counties.

In making this location Davis said he was guided by a deep faith in divine providence, and it was to divine guidance and not

to geology that he ever afterward attributed his success. The pre-
amble to the charter of the Luling Foundation (it is significant
that it was not called the Edgar B. Davis Foundation) begins:

Believing that a kind and gracious Providence, who guides the Desti-
nies of all humanity, directed me in the search for and the discovery
of oil, and in our successful management and favorable outcome
of the business, and believing that the wealth which has resulted has
not come through any virtue or ability of mine, and desiring to dis-
charge in some measure the trust which has been reposed in me; and
in a spirit of gratitude to the Giver of all good for his benefi-
cence....[33]

So successful was he in giving his money away that, although
the state of Massachusetts secured a judgment, its representatives
could find no assets to attach. The people of Caldwell and Hays
counties resented the attempted raid by a Yankee state. Not all of
them shared Davis' piety, but whatever their views on Divine
Providence, Davis was their hero.

A feature writer for the *New York Times,* in an article published
July 5, 1931, eight months after C. M. Joiner had brought in
the discovery well in East Texas, wrote, "Geologists and other
prospectors for the big oil companies had gone over this territory
looking for signs of oil, and had pronounced it barren, or at best
unfavorable for commercial development." The statement is not
wholly accurate, but it is what a journalist would have heard who
visited East Texas in 1931.

It is less misleading than one that appeared in Joiner's obituary
in the *Dallas News,* March 29, 1947:

C. M. (Dad) Joiner, 87, the Shakespeare-quoting wildcatter who
sank a battered bit into history's greatest oil field will be buried Saturday morning at Hillcrest Mausoleum.
Flat broke at 65, Joiner sank the epoch shattering East Texas field's
first well on a shoe string on land where geologists had already ruled
there was no oil.

The reporter was merely restating what had long been a popular tradition.

Probing for oil in the five counties (Upshur, Gregg, Rusk, Smith, and Cherokee) in which the East Texas Pool lies began as early as 1901. Among the men who drilled was A. P. Boynton, who put down three wells, one of them only six miles from the pool Joiner was to discover.[34] Further interest was stimulated by the Mexia discovery in 1921. In that year J. A. Colliton drilled five wells in Cherokee County, having in the latter part of his venture financial help from Colonel Humphreys, the discoverer of Mexia. In one of the wells he struck oil, but failed to obtain commercial production. He attempted to interest the major companies, and according to local tradition, was told by the representative of one of them that he would drink all the oil produced in Cherokee County.[35]

Whether this is true or not, the major companies did send their geologists into the region. They were looking for faults and salt domes, which they did not find, and most of the major companies withdrew. But the vote of no-confidence was not unanimous. Two geologists, Julius Fohs and James H. Gardner, had in 1915 recommended drilling near Kilgore. Their client had interested a subsidiary of the Shell Company in the project. The president of the company, also a geologist, drilled a mile from the location recommended. In so doing he missed discovering the East Texas field.[36]

Mose Knebel, then geologist for the Humble Company, induced his company to take leases. He failed to persuade them to drill, but the leases gave Humble a considerable interest in the East Texas field. Albert E. Oldham, a geologist for the Amerada Corporation, recommended that his company take leases on a block eight miles wide and twenty miles long at a cost of $150,000. The management was unwilling to commit this sum. If it had, Amerada would have been the biggest producer in East Texas.[37]

Finally, there is the seldom-mentioned fact that Joiner availed himself of geological advice, though the extent to which he followed it is not clear. Joiner had been a client of A. D. Lloyd in Oklahoma,[38] and after he acquired his East Texas leases, he postponed making a location until Lloyd, then busy in New Mexico, could advise him. But because he had to agree to drill on the Daisy Bradford farm in order to obtain a lease on it, he set up his rig two miles from the spot Lloyd had picked. Whether Lloyd approved this second location is not clear. After the loss of two holes, it was necessary to move to a third location. Mrs. Bradford and the crew moved the rig down hill, the path of least resistance. Even so, a sill broke after they had gone about 250 feet, and there drilling began on the Daisy Bradford No. 3. Lloyd did not approve of this location. He left East Texas and did not return until the oil sand had been reached. When the well came in, he and Joiner posed for a picture, and he was quoted as saying, "Boys, this is the fourth time Joiner has found pay sand upon my recommendation, and we're not going to let it get away from him this time."[39] Later he went before the Henderson Chamber of Commerce and pleaded with Joiner's creditors to refrain from filing suits and to give the discoverer time to develop his properties.

There was then a minority report on East Texas. The drill proved that the majority was wrong, and "Remember East Texas" became among geologists a warning against dogmatic inflexibility. But the companies that had been misadvised did not fire their geologists. The skeptics had already been converted.

J. C. Donnell of the Ohio Oil Company had once said, "When geology comes into the oil industry, I go out." Yet in a few years the geology department in his company was regarded as the most important.[40] The Gulf Oil Corporation in 1911 hired M. J. Munn away from the United States Geological Survey and directed him to organize a geological staff.[41] By 1920 all but one or two of the large companies had followed suit. The record of geology had been impressive. It had led to the finding of three-fourths of the oil discovered from 1920 to 1929.[42] During the next decade the refinement of geophysical instruments would give geologists and geophysicists new methods of exploring subsurface geology and further increase their chances of finding oil. Half the new reserves discovered from 1930 to 1959 were found by the use of the reflection seismograph.[48] Yet no instrument can find oil directly. Oil is still where you find it, and the only infallible geologist is, as John Galey said sixty years ago, Dr. Drill.

9

The Oil Promoter

ONE OF THE MOST VIVID of the folk characters of the oil industry, one of the earliest created, and one whose ancestry goes back farthest in time is the oil promoter. For he corresponds to the trickster, known in every culture, literate and preliterate, as one who lives by his wits, his only weapon being cleverness, his only technique deceit. He is everywhere regarded with a mixture of admiration and condemnation. His universal appeal is not easy to account for. It may be assumed that he is in some sense a projection of a quality inherent in the human condition: perhaps of our consciousness of our insufficiencies in our universe;[1] of our intuitive knowledge, if not conscious awareness, of original sin, expressed in the adage that "there is a little larceny in us all"; or of revolt against necessary social restraint.

Sometimes the trickster will have our complete sympathy, sometimes our bitter contempt; we may rejoice to see the trickster tricked. Our attitude will depend partly upon the motivation of the trickster, partly on our sympathy or lack of it for his victim.

The trickster no longer sells money-bearing trees to simple country folk, not even gold bricks to country merchants. But the rise of industry, of finance capitalism, and more recently of mass communications, has opened doors of opportunity that would have bewildered Pedro Urdemales or Sam Slick. There is no evidence that the trickster is disappearing from our culture and our lore. Now, in his most sophisticated persona, he manipulates symbols of popular value from his Madison Avenue office. But for the first thirty-five years of this century he was most often and most conspicuously an oil promoter.

In a denotative sense, anybody who promotes an oil venture is an oil promoter, but among the oil folk he is one who promotes a venture from which he hopes to gain whether oil is found or not. Whether he sells interests in a well or stock in a company, he expects to be compensated for his trouble, oil or no oil. This is not to say that he always, or indeed in most instances, is a trickster. There were and are legitimate and ethical means of promotion. For a simplified example often used in the past, let us say a promoter secures a lease on a town lot in Breckenridge in 1918. He estimates that he can sink a well to the producing sand for $50,000. He sells seventy-five one-per-cent interests in the well for $1,000 each. He has $25,000 above the expected cost of the well, and if he finds oil, one-fourth of the seven-eighths remaining after the landowner's royalty has been deducted belongs to him. And he hasn't deceived anybody.

He becomes a trickster if he does what many promoters are reported to have done—that is, if he sell interests totaling more than 100 per cent. I have been told that two men operating under this plan had the misfortune to strike oil and that they plugged up their wells and left in a hurry.

There is no conclusive reason for assuming that these latter men are more typical of promotion than the first-mentioned. Nevertheless, they and their kind have cast suspicion upon the whole fraternity. It is significant that Dad Joiner's friends have resented his being called a promoter. They want it understood that he was a bona fide wildcatter who was seriously engaged in the occupation of looking for oil.

This suspicion dates from the first oil boom, when the gold excitement had somewhat abated, and sharp practicers were turning to oil. In 1865 a music publisher, taking advantage of the excitement that Drake's well had started, brought out *The Oil on the Brain Songster*,[2] in which are several songs satirizing oil promoters. The verses of one such song consisted entirely of a listing of imaginary companies:

FAMOUS OIL FIRMS

By E. Pluribus Oilum

There's "Ketchum and Cheatum,"
And "Lure 'em and Beatum."
　And "Swindleum" all in a row;
Then "Coax 'em and Lead 'em,"
And "Leech 'em and Bleed 'em."
　And "Guzzle 'em, Sink 'em and Co."

There's "Gull 'em and Skinner,"
And "Gammon and Sinner,"
　"R. Askal and Oil and Son,"
With "Spongeum and Fleeceum,"
And "Strip 'em and Grease 'em,"
　And the "Take 'em in Brothers and Run."

There's "Watch 'em and Nab 'em,"
And "Knock 'em and Grab 'em,"
　And "Lather and Shave 'em well," too;
There's "Force 'em and Tie 'em,"
And "Pump 'em and Dry 'em,"
　And "Wheedle and Soap 'em" in view.

There's "Pare 'em and Core 'em,"
And "Grind 'em and Bore 'em,"
　And "Pinchum good, Scrapeum and Friend,"
With "Done 'em and Brown 'em,"
And "Finish and Drown 'em,"
　And thus I might go to the end.

Similar ridicule appeared in newspapers and periodicals. The Boston *Commercial* published a burlesque prospectus of The Munchausen Philosopher's Stone and Gull Creek Grand Consolidated Oil Company, with a capital stock of four billion dollars, and working capital of $39.50. Dividends were to be paid semi-daily, except on Sunday. The directors were S. W. Indle, R. Ascal, D. Faulter (treasurer), S. Teal, Oily Gammon, and John Law. In the same year (1865) the *Typographical Advertiser* announced the organization of the Antipodal Petroleum Company, with a capital stock of one billion dollars, with a par value of $10,000 per share, but offered to the public at twenty-five cents. The company proposed to drill through the earth and thus obtain production in both the United States and China from a single well. The treasurer was Mr. Particular Phitts, and the president was The Hon. Goentoem Strong.[3]

Thirty-five or more years later Texas was to have its quota of D. Faulters, and R. Ascals, and Particular Phitts. In referring to some of them I have used such terms as "it was said," "it was reported." Some of my informants have been reluctant to name names or to give clues to identification. Thus not all my information can be verified, nor does it need to be. I am concerned with the public image of the trickster-promoter. I shall not attempt to follow in detail the careers of any of the notorious tricksters exposed in court. It was their tricks that brought great crowds into the courtrooms, and that led to legendary embellishments of their exploits. It is the stereotype that commands our interest. There is, however, sufficient documentary evidence in the form of newspaper advertising, exposures in journals, and reports of court trials to suggest that the popular image had its objective correlative.

The trickster's greed is taken for granted, and a degree of cleverness, which, however, did not always see him through. How did he operate? What were his tricks?

The simplest of all was well-salting, a trick he might have learned from the gold and silver mine promoters that preceded him. The first salters of whom I have record were a couple of Vermont Yankees lured to Pennsylvania by the first oil boom. In 1864, when Alfred W. Smiley was working as a clerk, a report reached him that an abandoned well had been deepened and was good for twenty or thirty barrels. Smiley went to the well and saw oil flowing into a storage tank. The Vermonters sold the well to a Bostonian for $40,000 cash. The purchasers found that the pumps had been rigged so that oil from the tank flowed back into the well to be pumped back into the tank in an endless cycle. But before this discovery was made the gentlemen from Vermont who had made the sale were "extremely absent." Apparently they were never brought to trial.[4] Thus the country boys triumphed over the city slicker.

This trick, then, seems to have been a Yankee invention, but Texans have not been averse to making use of it. In 1921 John H. Wynne and his partner acquired six sections of land in Reeves County. They had not particularly wanted the land, but had taken it on a debt when money was not forthcoming. Six months later word came that oil had been found nearby, and a lease hound offered them five dollars an acre. This was more than they had considered the land worth in fee simple. But instead of signing up, they decided to investigate. Wynne found the well, and a man on duty who would open a valve, whereupon oil would flow from the casing head. It would be permitted to flow only a few minutes, for, the operator explained, he had no storage. But the driller gave the secret away. The oil flowed from a tank car on the railroad siding nearby, and there was enough gas in the well to bring it to the surface. Leasing activity ceased abruptly and Wynne and his partner did not get their $19,200.[5]

This method of salting is rather crude. The more sophisticated salter leaves the evidence for others to discover or interpret. This is more convincing and less dangerous. He may sprinkle the derrick with crude oil, and if he means to sell stock, he will photograph it. He may pour oil in the slush pit, or bring oil sand from a producing well and leave it on the derrick floor for a scout to find. When Dad Joiner's driller reached the oil sand, he washed the bit in a bucket of water and left it on the derrick floor. The stories resulting from this are too many to recount here. One man told me of going to the derrick and finding it unattended, of examining the sand, and finding all the evidence of salting. He joined the legion of East Texans who failed to grasp the forelock that Dad Joiner's discovery provided.

It was inevitable that some well salter would be hoisted with his own petard, for wherever there are stories of tricksters, there are stories of the trickster tricked. My story comes from Burk-burnett and the operator will be called A. D. Siever. He poured crude into his well and hauled it out with a bailer in the presence of prospective buyers, a couple of New York Jews. They bought, and Siever was happy in his success. To outwit a New Yorker was a considerable achievement, to outwit a Jew was a greater achievement, but to outwit two New Yorkers and two Jews at one and the same time was a superb achievement. But Siever's complacency was short-lived. The New Yorkers deepened the well and brought in a producer worth many times what they had paid for it. And Siever never knew whether his salting sold the well, or whether the buyers had geological information unknown to him.[6]

How extensively well salting was practiced can hardly be known, but both the legal record and the oral tradition would indicate that it played a relatively minor role in dishonest oil promotion. The trickster relied mainly upon the sale of corporate stocks, and his indispensable assets were imagination and verbal skill. Mr. Ecks furnishes an example of what was needed. Hauled into court on a charge of selling fraudulent oil stock, he somehow got access to the witnesses for the prosecution. He sold every one of them a share or more of stock, and agreed to take in payment their vouchers for mileage and witness fees. The judge declared a mistrial. Years later my informant saw Mr. Ecks, who told him that he had suffered his punishment, had repented of his sins and had been converted to religion, and was now successfully using his talent in the service of the Lord. He was a revivalist.[7]

My informant was silent upon the name of Mr. Ecks's company, but I am sure it was well chosen. For the trickster knew that a rose by any other name would not smell so sweet to the people on his sucker list.

If a company was a modest one, with holdings limited to a single locality, the trickster often sought a name that would link his company with a producing well or field. For example, the discovery well at Desdemona was on the Duke farm. Boyce House has listed eighteen companies making use of the word *Duke:* The Grand Duke Producing Company, Heart of Duke, Duke Extension, El Paso Duke, Italian Duke, Post Duke, Duke-Burk-Ranger (three fields represented), Duke Knowles Annex, Royal Duke, Duke Consolidated Royalty Syndicate, Erath Duke, Duke Dome, Alma-Duke, Tex-Duke, Giant Duke, Duke of Dublin, Comanche Duke, Iowa Duke.[8] Some of the Dukes were found to be fraudulent, as was also the Blue Bird Oil Company, one of several making use of symbols of luck. Others were Lucky Boy, Lucky Seven, and Rainbow. Uncle Sam suggests patriotism, but the company so named was a notorious swindle. The great supercorporation through which Dr. Frederick A. Cook and Seymour E. J. Cox swindled thousands bore the innocent and co-operative-sounding title of The Petroleum Producers Association.

Another device was to choose a name suggesting an affiliation between your company and a well-known successful one. Two such companies offering stock for sale during the Beaumont boom were the Rockefeller Oil Company of Beaumont and the Stephenville Standard Oil Company of Beaumont. The only connection they had with John D. Rockefeller or any of his Standard companies was in the names.[9]

The General Lee Development Interests would seem at first sight an inspiration for a corporation seeking to sell stock in Dixie. But the organizers were not content to rely upon the magic of the name. They found a janitor named Robert A. Lee, conferred a military title upon him, declared him a descendant of Robert E. Lee, paid him $12.50 a day for the use of his name, and described him as a famous geologist. Their literature proclaimed that just as "Robert E. Lee gave his life to the South, so is now General Robert A. Lee giving his life to the oil industry and the cause of humanity." But a federal court found that he was no kin to Marse Robert, and that he was not a geologist. Nor was the jury convinced that he was giving his life to the cause of humanity. He and sponsors paid fines and went to a federal prison.[10]

During the heyday of the oil trickster, from 1918 to 1924, radio and television were not available, but he had other means of getting his message to the public. One was to get his propaganda published in reputable papers as news. A notorious example is a story that appeared in the New York *Sun* in 1903 (May 10). Whether payola was involved, my sources do not reveal.

Under the 200 square miles of rolling prairie land controlled by Mr. King and his associates, there is a vast sea of petroleum. While its length and breadth have been pretty well established, no plummet has ever yet sounded its depths.

It lies in its subterranean bed, where it will sleep until the suction pumps of the big King-Crowther Corporation begin to thud and clank in the oil-filled caves. As yet, the surface has been barely scratched, as it were, and seven oil wells have been found. By a fair process of reason, it may be assumed that in the entire 200 square miles of territory, when fully developed, there should be at least 8000 oil wells.

The general idea is to pay the investor not less than 20% a year so that in five years he will receive his original investment, leaving a profit of from three to five times the original amount.

These estimates are based on Mr. King's knowledge and experience. As a matter of fact, investments may pay anywhere from three to five times in excess of the figures quoted.[11]

A few days after this story appeared, the attorney general of Texas filed a petition alleging that the King-Crowther Corporation had obtained its charter through false information. The court estimated that shareholders had been fleeced out of two million dollars.

As newspapers grew more wary of publishing free advertising, the trickster, knowing the advantage of the seemingly impartial news story, began acquiring control of oil journals by founding, by purchase, or by lending them money. The *World's Work,* in an article exposing fraud in the oil industry (1923),[12] listed eleven periodicals as owned or controlled by oil promoters: *Pat Morris Oil News (Fearless and Truthful Oil News), Independent Oil and Financial Reporter (Fair, Faithful and Fearless), International Investors Bulletin, Independent Oil News, Texas Oil World, Texas Oil Ledger, National Oil Journal, Arkansas Oil and Mineral News, The Banker, Merchant, and Manufacturer, Mining and the Industrial Age,* and *Commercial and Financial World.* It was charged that one of these was for a time hostile to Dr. Cook, but was brought to see the light by a substantial loan.

But not all tricksters could afford an independent oil journal. They relied chiefly on newspaper advertising and direct mailing. In spite of the better business bureaus and the Vigilance Committee of the Associated Advertising Clubs of the World,[13] which in 1921 declared that 95 per cent of oil stock offered through newspapers was unworthy, it was still possible to get wide coverage. Because security laws were more lax than now and because both the Post Office and Attorney General's Departments were grossly under-staffed, fear of legal action was not a wholly effective deterrent, a fact which gave vigor and scope to the imagination.

The *New York Times* (March 9, 1924), commenting particularly on conditions in Texas and more specifically in Fort Worth, known as the capital of fraudulent oil promotion, observed that the oil stock promoter

has contributed some of the glibbest, most convincing writing of our era. Some of this writing is so broad and highly colored that the secret story should be plain to see. But the public is greedy for this sort of fiction, and the oil stock frauds flourish in the fertile soil of the public imagination.

A sampling of this writing tends to support the judgment of the *Times* editorialist. There were a number of approaches, each with its appropriate style and tone. One might be called the unqualified promise, delivered in bold, simple, and direct English:

Oil will always be in demand, it will always yield its fixed, profitable price, and it will make dividends to the holders of shares in the Bonnabel Refining Company as sure and steady as the progression of time. . . . The company can guarantee at least 20% per annum.[14]

I absolutely guarantee $200 returned for every $100 loaned me for oil development, with a full 100% profit remaining as a permanent investment.[15]

"Conservative estimate of profits . . . 350%." On the inside cover of the prospectus was the following unacknowledged quotation:

> Our doubts are traitors,
> They make us lose the good
> We might win by fearing to attempt.[16]

One syndicate "aimed at" 10,000 per cent profit in from four to six months.[17]

Such statements as these are what the *Times* writer had in mind as so broad and highly colored as to give the secret away. One could sound more convincing by promising big profits and then adding qualifications that would leave the profits still big. One promoter had a tract of a hundred acres upon which he said there was room for fifty wells.

Even if these fifty wells should make *only* 100 barrels apiece, the oil, being Pennsylvania grade, commands a price of $4 per barrel at the well, and will net us a nice profit . . . 50 times 100 will be 5,000 barrels a day, which at $4 per barrel, will equal $20,000 a day. This multiplied by 300 [perhaps the wells were not to produce on Sundays and holidays] will make a total income of $6,000,000 a year from this 100 acres alone.

But he goes on to say that 10 wells would be sufficient to develop the property, and, even if the production were only 25 barrels per well, the income would still be $300,000 a year.[18]

Another theme is: others have got rich, why not you? Often

there was the warning that you had better hurry, for the stock
was going up!

ARKANSAS GIRL MAKES $300,000 ON OIL ACREAGE. BUYS LAND FOR
$500; REALIZES $900,000; BIG PROFIT, EH, WHAT? THREE MEN
POOLED THEIR ALL, $25. AND SOLD OUT FOR $250,000.[19]

Others preferred a tone of frankness. There is risk in the oil
business:

If you ask my advice about investing, I don't give any. I have acted
on my own judgement, and have invested in the Company with which
I am identified. I don't advise anybody either one way or the other. If we
strike oil our stock will be worth ten to one, or more. If we don't
strike a well there will be no difficulty in selling our holdings at a
greatly advanced price, as values are doubling, quadrupling and
quintupling there from day to day. I determined, however, that I
would give nobody advice in the matter. If I should give the advice
and our stockholders made a thousand per cent on the investment,
they would think I was a great man; but if I gave the advice, and
there was nothing made out of the investment, they would lay the
failure in the realization of their fast profits, to my account. I shall
accumulate no such liabilities as these. I have unbounded faith in the
oil fields there, and I believe they are going to supply the fuel and the
illumination for the world; but I don't advise either one way or the
other. I can only say this: If any readers... are going to buy oil stock,
our oil stock is as good as any oil stock in the market and far better
than ninety per cent that is being sold.[20]

If you cannot afford to take a chance to lose $10 to $25, do not get
into this scheme because oil investments are uncertain. Admittedly it
is a "long shot." The backers of the company have such a firm belief
in it, based on the best geological information obtainable, that they
have put their own money into it.[21]

In our exaltation of the hero we have sometimes without the
warrant of fact made every discovery somebody's folly. There were
Columbus' folly, Fulton's folly, Morse's folly, and in oil Drake's

folly and Higgins' folly, and if we may trust a newspaper advertisement, Carruth's folly. One of his ads ran:

Hog Creek Carruth: The name that will live throughout the ages as the name of the man who toiled singlehanded for seven long years to prove up his belief and attain his goal—who traced an oil structure 20 miles across the ranges from Strawn to Desdemona—who conceived and organized the famous Hog Creek Oil Co.—who drilled the discovery well of the great Desdemona field at one time called the richest spot on earth—who transformed the desert into a fountain of liquid gold—who built a city of 30,000 souls from a village of 200 people and who paid every person who held shares of stock in his renowned Hog Creek Co. $10,133 for every $100 invested.[22]

Thus were the doubting Thomases confounded.

One very successful copywriter found the folksy style effective:

Now, folks, I'm going to tell you a lot of things in this ad that will be good for your souls. I'm not a promoter and I'm not an ad writer—I'm just plain old Harry Bleam. All I can do is just sit down and tell you this stuff the way I know it to be. Most everybody knows about Harry Bleam. I'm just a common guy. I'm not in politics, but I'm a square shooter from who laid the chunk, and I put my cards down on the table face up. I'm not a promoter. You will know that anyway in just a minute, because I'm going to tell you something no promoter ever told you. If the well I'm getting ready to drill on my big little 4-acre tract down there don't come in a gusher, then there ain't no such thing as a bellyache.... Why, if I didn't believe I was going to get oil here—and I don't mean just a dinky little 1,000 barrel well—I mean a 10,000 or 15,000 barrel well—I'd never try to sell anybody an interest in it....[23]

Boyce House tells of a promoter whose literature showed a picture of a cell-block at the Leavenworth Prison underneath which was the statement, "The doors of this prison will open to receive .... if he fails to make good every statement made to the public."[24] This statement is no mean achievement. But Dr. Cook could visualize a worse fate. Here he describes his thoughts while watching an oil-well fire.

I stood on a hill about a half a mile away watching this shaft of light and heat as its wicked tongues of flame leaped and roared while the men rushed around in their feverish haste to extinguish this great torch of the oil fields. I was standing there with the black of the night behind me and the clear white light of this burning well in front of me, wondering if possibly all this roaring fire wasn't in reality sent as a kind of warning to the fake promoters—the meanest rodents that ever breathed God's pure air.

I don't believe that a man who would wilfully defraud the public and take from the investors who are willing to help develop nature's resources the money which they have so carefully saved, without giving them a fair return, deserves much better an end than might be typified by this flaming gas well.

The preachers tell me that the day of fire and brimstone in the church is past but we have plenty of it left here in the oil fields and I wonder after all if his Satanic Majesty isn't retaining just a little supply of the old fashioned hell-fire torment for the reception of a few phony promoters.[25]

Whatever the ultimate fate of Doctor Cook, a cell block at Leavenworth did open to receive him.

What the promoter-trickster's take was can only be hinted at.

Within eight months after the Spindletop discovery, the capitalization of Texas oil companies had reached $231,000,000, although actual investment in the Beaumont field was estimated at only $11,000,000.[26] Within a year there were five hundred Texas oil companies doing business in Beaumont, not to mention hundreds more chartered in other states.[27] Not wholly untypical were four companies capitalized at one million each. Their assets were a jointly held lease on a block of land forty-five feet square.[28] It is little wonder that Spindletop became known as Swindletop.

Carl Coke Rister says that out of 1,050 new stock companies formed in 1918-19, only seven paid dividends. The Oklahoma

Commission of Corporations estimated that only one dollar was returned out of every $550 invested in stock companies. In Kansas it was estimated that in 1916-17 only 12 of 1,500 new companies showed profits. An Oklahoma writer estimated that there had been a $555 capitalization for every barrel of oil produced.[29] In 1924 the *Financial World* estimated that the capital of defunct oil companies had aggregated $500,000,000.[30] Rister estimates a total accumulated investment of $102,000,000,000 up to 1947, against a return of $61,000,000,000.[31]

It is not to be assumed that all, or even a majority, of the unsuccessful companies were fraudulent. But there are indications that the aggregate sum that went into the pockets of the tricksters was considerable. The *World's Work* in 1918 said that one promoter (he was jailed) had fleeced 25,000 people out of $2,500,-000.[32] In February, 1924, there were in federal courts sixty-three cases pending against persons representing or claiming to represent Texas oil companies, who, the Department of Justice claimed, had taken in $140,000,000. A year earlier the solicitor general of the Post Office Department estimated that during the preceding five years $100,000,000 had been lost to fraudulent promoters in Texas alone:

No doubt some of these companies were started by men who hoped to strike oil and make money from the production, but in practically every case the promoters had laid their plans to profit from stock selling, regardless of the result of field operations. Seldom was it that a promoter invested his own money.[33]

The great crackdown came in 1923. The Securities and Exchange laws of the 1930's made it considerably more difficult for the trickster to operate. I would not be so bold as to say that he has left the oil industry altogether, but his heyday in this industry is over.

A convicted bank robber once made this defense of his trade: When he robbed a bank, nobody lost a dollar but the insurance company, and it obviously gained in the long run, for if bank robbing ceased the sale of robbery insurance would cease soon afterward.

I know of no such ingenious defense's having been made of the oil trickster. When he was brought to trial, he typically pled guilty. When he pled not guilty, he was an honest man who had exposed himself to the great hazards of the business in a sincere effort to help his fellow-man. If, in imitation of the bank robber, he had attempted to rationalize his conduct, he might have said that his social role was to alleviate the disastrous economic effects of oversaving. But Keynesian economics were not in fashion in the days of Harding and Coolidge and Hoover.

10

The Shooter

THE OIL INDUSTRY produced no more spectacular character than the shooter, and of the shooters the most legendary was Tex Thornton.

Before the process of acidization — that is, putting acid into a well to dissolve the lime and thus increase the porosity of the formation — was developed, charges of nitroglycerin were often detonated to break up tight formations and "jug out" the bottom of the well. The men who specialized in this work, some of whom also fought oil and gas well fires, were called shooters.

These men and their work and the explosive they used were subjects of a lively if somewhat morbid popular interest. In those prenuclear days nitroglycerin was regarded as the most powerful of explosives. A twenty-eight-pound blow or a temperature of from 118 to 120 degrees Fahrenheit was sufficient to set it off. As it could not be shipped by rail, the ingredients were mixed locally, and sometimes inexpertly, and the compound was transported to the wells in specially designed containers first in spring wagons and later in cars. Accidents occurred in factories, on the roads, and at the wells, and they were reported in gruesome detail, as the following dated examples will show:

1867: "They didn't find enough pieces to fill a cigar box," was the way they used to close the story of petroleum's first nitroglycerin fatality. William Munson had a factory near Reno for making nitroglycerin. He was working in this factory one day in 1867 when an explosion—there were no witnesses to tell how it occurred—blew him and the building to fragments.[1]

106

1871: Parts of the face, with mustache and four teeth attached, was the largest portion of the driver recovered from the debris. The horse was disemboweled, and to numerous trees lots of flesh and clothing were sticking. From the ghastly spectacle the beholders turned away.[2]

1919: Two men were blown to pieces, a woman slightly hurt, a bridge demolished, and a garage badly damaged when a cargo of nitroglycerin being transported in an automobile blew up on the Weatherford road about twenty miles west of Fort Worth at 7:45 this morning. One of the persons killed was the driver and the other was a passenger he had picked up. Neither of the men has been identified.

Both men were blown to atoms. Human fragments were picked up along the road and in the fields for a distance of several hundred yards from the scene of the explosion.

Pieces of the car were scattered to the four winds. The left front wheel, with only the nubs of the spokes around the hub, was lying 350 yards away.

In the edge of the ditch skirting the roadway probably 600 ft. from where the mishap occurred, searchers found the exterior skin of a man's face. It looked as if it had been removed with a knife and resembled a mask made of human flesh.[3]

1927: As a result of the first accident in the shooting of a hundred wells in the Panhandle oil field, Paul Wright, of Amarillo, and Wayne Williams, tool dresser at the Empire well near Pampa, are dead, and Robert Cheatwood, another tool dresser, is seriously, if not fatally injured. The tragedy occurred when a bomb exploded as Wright was lighting it to drop it in the Empire well where 500 quarts of nitroglycerin were to be exploded.

Mr. Wright's hands, it was said, were found blown away yards from the scene, but otherwise his body remained intact.[4]

When the shooter arrived at the well to be shot, he first suspended a long can, resembling stovepipe, over the well and filled it with nitroglycerin, being careful not to spill a drop, for a drop spilled on the derrick floor could cause a fatal accident. Then he would lower the shell on a small cable, in later times made of steel wire. The hook that engaged the bail of the shell was designed to release the shell when it rested on the bottom of the well. For large shots several shells were lowered one at a time. Several kinds of detonators were used: bombs with time fuses, electrically activated caps, and time bombs with watch mechanisms. After the well was loaded, it would be partly filled with water or oil to keep the explosive force from expending itself upward.

What the shooter dreaded most of all was an irregular flow of gas by "heads." If these heads came with the regularity of Old Faithful, they would present little danger. But they did not always behave like that. The shooter would time the interval between heads and in most instances be safe. But sometimes a slack line would show that an unexpected head was sending the charge back up the well. When this happened, the shooter might run, hoping to get behind a barrier or to get three hundred feet or more away before the shell came out, hit something, usually the crownblock of the derrick, and exploded. Or he could remain at the casing head and attempt to catch the shell as it came out, knowing that if he failed he had little chance of survival. There were several reasons why he might fail. The shell would be covered with water or oil, in either case hard to hold. The shooter might be blinded, or even overcome by the gas. Once caught, the shell had to be handled with the utmost gentleness; and if a time bomb had been attached it had to be removed quickly and disposed of.

In view of these hazards, stories of shell-catching may well be received with some degree of skepticism. D. D. Kling, writing in 1932, said that "incidents of this kind are exceedingly rare. In fact, although the average shooter will claim to have performed the feat, there is only one authenticated instance of this nature on record."[5] Kling's skepticism, however, is excessive. The incident he refers to is probably one that happened in Ohio in 1895:

The shooter, McCoy, felt the line slack, knew what was happening, raced to the mouth of the well, knelt down, and, as the shell emerged he caught and clasped it to his body with his right hand and pressed down on it with his left hand, meantime he was drenched with the flow of oil.[6]

Besides the unidentified shooter to be mentioned later, at least four men had caught shells in Texas before 1932. Jack Rapp, in the Ranger field, waited at the casing head with an ax handle in his hand, and when the top of the shell emerged, he thrust it through the bail and held it until the flow of gas subsided.[7] In 1927 Tex Thornton, at Dixon's Creek in the Borger field, caught a shell in his arms.[8] And in North Texas, sometime before 1918, Tom Mendenhall had done the same thing. Walter Cline has given an eyewitness account:

Tom Mendenhall came out to shoot it [a well Cline had drilled], and there was quite a bit of gas in the well; and we loaded a six-foot bucket with liquid nitroglycerin and were letting it down in the hole.... The ... bucket ... I guess squeezed the gas a little bit and it made a blanket or a pocket in there, and when we let the second bucket down, the gas just caught it and began to push it back up.... Well of course the only thing I could think of was the tall uncut timber. I never was a speed demon, but I figured I'd try to make a new track record. And Tom was standing right by the well. And he put his head down and he could hear the glycerin bucket scraping on the side of the casing as it came up, and he said, "Don't run off, Walter. I'll take care of it." And so foolish me, I stood there to see what he was going to do, and he did something I wouldn't do at all. He just spread his legs on each side of the casing and waited until the gas pushed this bucket of liquid nitroglycerin right up. And when it got up about waist high, he just reached out and hugged it like he was coming home to mama, and picked it up the rest of the way and took it over in the corner of the derrick and set it down.[9]

Evidently the shooter by listening could estimate with some accuracy the speed of the returning shell and thus his chances of a successful catch. I know of only one attempt that failed. Powell

Wright let the shell slip through his arms. It struck the crown-block in the top of the derrick and exploded. Wright had time to jump behind the engine house, and though he was knocked unconscious, he was not seriously hurt.[10]

In the Caddo field not far from Ranger, a shooter by the name of Davidson of the Independent Torpedo Company came to shoot a well. The well had been flowing at forty-five-minute intervals, but with the lapse of that time, it remained quiet. After waiting three hours, Davidson decided that the well was dead and prepared to shoot. He had hardly lowered the thirty-quart shell when the slack line told him it was returning. He caught the shell at the top as it emerged, but it was too long for him to lift from the well. He lifted it as high as he could, and let the nitroglycerin run on the derrick floor. With the load lightened and with the bent over portion to hold to, he was able to get the shell out of the casing. His only injury was to his hand.

He put his whole weight on it [my informant said] with his hand clenched over the sharp edge of the shell.... His fingers were cut in a row. He was showing us this place where he cut his hand. His hand was just as still as mine is right now; he wasn't shaking.[11]

The question naturally arose as to what kind of man one must be to choose to make his living handling this dangerous stuff called nitroglycerin. There were two traditional answers. He was a calm, competent, and modest expert who went about his business with no more thought of heroics than a shoe clerk or a bank teller. Whatever he did was all in the day's work. Or he was a devil-may-care kind of fellow, whose chief pleasure in life was flirting with death. There was some basis in reality for both stereotypes. To the first would belong Jack Rapp, Tom Mendenhall, and the unidentified hero of Boyce House's "Sunday in Breckenridge."

The well in this account was close to a small church. The congregation had assembled. As the first shell was being lowered, came the slack line and the sound of gas. The shooter told the other workers to run. He was about to follow them when he heard the strains of "Jesus, Lover of My Soul" coming from the church. He returned to the casing head and braced himself, and caught the shell. The church service ended and the worshipers filed out not "knowing that the wings of the death angel had but a moment before fanned their cheeks."

"History has not preserved the name of the shooter.

Anyway, praise would only have embarrassed him."[12]

The other stereotype made its appearance in fiction in 1908, in a story entitled "Jim Hanks — Oil Shooter," by T. P. Byron. Before introducing his character individually the author makes this comment on the type:

> After a "shooter" has taken his last reckless drive, his friends mournfully collect him in a ten-quart pail. Sometimes they have to pluck him from the trees.
>
> "Shooters" are hard workers and hard drinkers, their peculiarities few but so pronounced that the class is always recognizable. They always drive fractious horses; they prefer rough roads or no roads at all...and dread nitroglycerin is a plaything to be handled with pleasure and satisfaction.[13]

Leave out the hard drinking and substitute a high-powered automobile with a specially designed body for the light wagon and fractious horses, and you have a reasonably accurate description of Tex Thornton, shooter and firefighter who attained legendary status during his lifetime. He was once quoted as saying that it was "the thrill of the game" that kept him in business,[14] a statement borne out by his own account of his boyhood.

He was born Word Thornton in Mississippi in 1891, the son of a physician. When he was nine years old, his father took him to a Fourth-of-July picnic at Tacopola, where the boy for the first time saw fireworks for sale. He spent all his money for firecrackers and begged his father for more. He came home with his mind made up that if there was a way to make a living shooting off explosives, that was the way he would do it. By the time he was sixteen, he had found a way. Farmers in Mississippi were learning about dynamite and were taking advantage of it as a cheap method of clearing their fields and pastures of stumps and reclaiming old cut-over timberlands. Contractors organized crews and did the blasting by the job. Word Thornton joined a crew as a helper. Within a few weeks he was bossing the crew, and in a few months he had given up his job and was in business for himself.

Then Dr. Thornton moved to Texas and opened an office at Goree in Knox County. He also bought a farm. There were no stumps to be blasted and Word had no liking for farm work. Neither did he have any desire to prepare himself for a medical career as his father wished him to. The oil industry was reaching into North Texas and young Thornton watched a shooter at work. He then let it be known that he was available for well shooting. Big pools were being opened at Ranger, Desdemona, Breckenridge, Burkburnett, Electra, and in 1926 the Texas Panhandle. He had little trouble in finding clients.[15]

It was as a fire fighter, however, rather than as a shooter that Thornton first became a popular hero. Who first thought of extinguishing an oil well fire by snuffing it out with an explosive, perhaps cannot be determined. A. H. Clough in *Petroleum Age* for February, 1920, gives the credit to a shooter named Ford Alexander. This was the famous Standard Oil well near Taft, California, which produced 180,000,000 cubic feet of gas a day, and which burned for ten days, generating heat above 300 degrees 250 feet from the fire. After the company had spent nearly half a million dollars and had exhausted the known methods of fire fighting, Alexander wrapped 150 pounds of blasting gelatin in asbestos and brought it into position by cable.[16]

But Frank Hamilton had seen a fire extinguished by explosives in the Caddo Lake field as early as 1908, under the direction of C. M. Chester, a superintendent of the Texas Company. The derrick had burned down. Chester stretched a cable over the well, upon which he sent into position a steel basket of dynamite to be exploded by a time bomb.[17]

And according to another informant, dynamite had been used successfully to put out a fire at Potrero del Llano, Vera Cruz, Mexico, in 1915.[18] Here the explosive was brought into position through a tunnel. And Tex Thornton is reported to have put out a fire with nitroglycerin at Electra, Texas, in March, 1919.[19] The owners of the well were skeptical and would not deal with him, but consented to let him try at his own risk. To protect himself from the flames he made what he called "a sort of Mother Hubbard affair" of asbestos. He later had a complete asbestos suit, including gloves, boots, and helmet, made by the Johns-Manville Company. If such equipment had been used before, he was not aware of it. Whether Ford Alexander preceded him in this is not clear.[20] A reporter for *Petroleum Age* (December 15, 1921) wrote:

Mr. Alexander, with the assistance of E. H. Clausen, a representative of the Johns-Manville Company, gave a demonstration of fire extinguishing never before attempted, and which is likely to stand for years as the most daring venture of its kind. . . . Alexander and Clausen, clothed in asbestos suits, carried the 80-pound charge wrapped in an asbestos covering, to the platform set up a few feet from the well.[21]

Whether or not Thornton was the first to use an asbestos suit, it became an essential item of his equipment, for his usual method was to enter the fire and place the charge by hand.

In ingenuity, versatility, and perhaps in oil and gas conserved, Thornton would rank below Myron Kinley, one of whose notable achievements was the putting out in 1931 of a well in Moreni, Rumania, that had been burning 890 days.[22] But a series of spectacular successes in the Texas Panhandle in 1927 and 1928 made good copy, and the journalists found Thornton not unwilling to co-operate. On May 28, 1927, at Sanford, near Borger, McMillan No. 1 caught fire. Eight men were burned to death and five others were seriously injured. As Thornton prepared to blast the fire, the newspapers reported his activities in great detail, and when he succeeded he was a hero. A few days later the Cockrell-McEllroy caught fire and claimed one life. Thornton shot out the flame. On June 9, the Empire Fuel and Gas No. 1 caught fire and Thornton was again successful. This time, when he entered the fire, he found the valve in working order and closed it. Thus within two weeks he had put out three fires, two of which had resulted in loss of life.

Then, on January 19, 1928, the Rachael No. 7 near Corpus Christi caught fire and cratered. Thornton was called. After he had removed the wrecked derrick and other metal from the crater and laid a pipeline to bring water from the Gulf of Mexico, he called his wife in Amarillo to bring a load of nitroglycerin. With a helper, she made the trip of eight hundred miles at an average speed of forty-five miles an hour. The Associated Press carried pictures of her and the special car she drove. Another widely circulated story and picture came from the scene. Harry Diehl, a news cameraman, borrowed an asbestos suit from Thornton to make some close-up shots of the fire. He was overcome by heat, and Thornton rushed in and carried him to safety. Another cameraman got a picture of Thornton carrying Diehl out. Thornton was now called the "human salamander." A member of the Texas Railroad Commission (which is charged with administering the oil and gas laws in Texas) declared that "Tex has more courage than Jack Dempsey, Gene Tunney and Tex Rickard all together."[23]

The *American Magazine* for March, 1928,[24] carried Magner White's illustrated article, "This Is No Job for a Nervous Man," in which appears for the first time the legend of the five shells.

Sometimes [Thornton is quoted as saying] the gas beats you to it, and starts coming up before you are ready for it.... That's what happened that time the five shells were blown up at me. I was just letting the fifth one down when I felt the line slacken and heard the gasolene engine speed up, because the load was taken off.

Some of the fellows started to run. They knew that if that shell ever hit the top of the derrick—good night! The gas was sending it up like it was a feather floating over a hot radiator, and the odor was almost enough to knock a man down.

I couldn't run. There was only one thing I could do. When that shell stuck her nose above the casin', I grabbed her. She was all covered with mud an' oil, naturally, an' it was like tryin' to catch a slippery fish.

I set her down as polite an' gentle as I could, and started to wipe the sweat off my face when along came another one. I grabbed that one too. I just had time to lay her back down, when the remaining three shot up, so close together that I was afraid they'd bump one another. By the time I laid the last one down, my helpers were all a hundred yards away, an' they'd done the distance in a little over ten seconds.

There are several reasons for questioning this story. The incident had not been reported before. Time and place are indefinite. Except in one story to be mentioned later, there is no other report of a shell's returning to the surface once it rested on the bottom of the well, where the gas pressure would be exerted mainly on the sides. The story differs in essential details from the version published in the *Kansas City Star* for November 9, 1930. Here the incident is reported as having occurred near Borger. The well was down 3,000 feet, and Thornton's plan was to explode three shells each containing thirty quarts of nitroglycerin.

I had lowered two shells [he is quoted as saying] slowly and carefully, of course, because a slight jar will sometimes explode it, and was lowering the third one. It was down nearly to the bottom, when I saw the line slacken and begin to come up. I knew instantly what had happened. Either gas or oil or both had started to rise in the pipe and was pushing those three shells ahead of it. In a few seconds they would be shot out of the mouth of the well casing unless I could catch them as they came out.

The drum on which the wire was wound was about sixty feet from the well. I ran for the well, yelling for everyone to run for their lives. I reached the top of the casing and straddled it, with my hands held ready to catch the first shell as it came out. If I missed one I was gone. The first shell emerged at a speed that would have shot it twenty or thirty feet and it certainly would have exploded as it hit something or would have gone off when it dropped. I grabbed it in both arms and gripped it with all my might. It was slippery with oil, and I tore my fingernail ends off both hands as I dug into it, but I held it.

It was eight feet long and the thirty quarts of nitroglycerin made it weigh ninety pounds. Clasping it in my arms I eased it up out of the casing until the top of the next shell showed. I put my foot on that second shell and held it down, while I gently let the first shell slide into the pool of oil that almost filled the celler of the well. Then I got rid of shell No. 2 and No. 3 the same way. Since then I've caught a dozen or more shells coming out of oil wells in the same manner.

If Tex was inclined to stretch the blanket a little, he had precedent for it. An old shooter by the name of William Meyers of Bradford, Pennsylvania, had a long time before regaled his listeners with the following story:

It was a close call, and no mistake. In the magazine I got some glycerin on my boots. Soon after coming out I stamped my heel on a stone and the first thing I knew I was sailing heavenward. When I alighted I struck squarely on my other heel and began a second ascension. Somehow I came down without much injury, except a bruised feeling that wore off in a week or two. You see the glycerin stuck to my boot heels and when it hit a hard substance it went off quicker than Old Nick could singe a kiln-dried sinner. What'll you have, boys?[25]

Thornton's fame led to an incident in Edna Ferber's *Cimarron* as improbable as any shooter's tall tale. When she was in Oklahoma and the Texas Panhandle gathering material for the novel, she heard of Thornton's shell catching. She had a character who needed a sacrificial death. She let him catch a shell some distance away from the well just as a football player might take the ball on the kickoff. The man was killed by the weight of the shell, but he saved the lives of others. It was a heroic act, but obviously impossible, since in the first place the shell would have been sure to hit the derrick, and in the second place even a small shell would have been too big to catch in this manner.

This event was in no way prophetic of the death of Tex Thornton. In 1949, after more than thirty years of shooting and firefighting, he was killed by a hitchhiker to whom he had given a ride.

11

The Driller

THE TRADITIONAL IMAGE of the driller — an image that emerged early and persisted late — was that of a footloose and fancy free wanderer, physically tough and indifferent to danger, competent, and in the knowledge of his competence proud and independent, often to the point of arrogance, yet a man of few words; a prodigious consumer of hard liquor, and a fighter from who laid the chunk. That not every driller, not even the majority of them, conformed to this image in detail did not prevent its development and persistence, and it does embody the qualities that receive most attention when veteran oil folk talk about drillers of times past.

Some of these qualities were dictated by the conditions under which the driller worked, and these conditions made the oil fields especially attractive to young men from farms and villages, from which a great majority of the workers came. They sought adventure, good wages, and in many instances an opportunity to get into a business that might, and often did, lead to wealth. The oil industry came at an opportune time for these men, for as the introduction of farm machinery created a surplus of farm labor, the oil industry was continually expanding. The country boys, who often had had experience with cotton gins, cottonseed processing plants, or sawmills, as well as with agricultural implements, found oil field work to their liking. It was outdoors and largely manual. They were accustomed to hard work and long hours and did not find the twelve-hour day and the seven-day week excessive.

They accepted risk as a matter of course. In the absence of adequate workman's compensation laws, a worker might or might not be taken care of in case of accident, depending upon his employer. Some employers had personnel policies ahead of their times, but safety devices were slow in developing. In the first years of this century the derrick man stood on a narrow board and worked without a safety belt. When belts were introduced some men refused to wear them because they were more afraid of fire than of falling. On rotary rigs the chain that drove the rotary was made of cast iron and was without housing. Sometimes it would break and wrap around a man's leg. There were no steel helmets to protect the head or steel-toed shoes to protect the feet. Without blowout preventers men were often gassed, and fires were too common. Early workers took all this in their stride, and thought no more about safety than their employers did. Indeed, the danger they were exposed to was a source of pride. They considered themselves men in a sense that counter-jumpers and pencil-pushers were not. But they knew the necessity for alertness.

They were wanderers by necessity. A wildcat well was drilled, opening a new field. Hundreds, even thousands, of men were employed. In a year or two the field was "drilled out"; that is, its limits had been defined and all the wells needed had been drilled. Storage tanks had been erected and pipelines laid. The number of men needed to maintain the field was a small fraction of those needed to develop it. Nearly any veteran driller could be cited to illustrate this mobility. O. G. Lawson will serve as well as any. When he was eleven years old, his father left the West Virginia hillside farm upon which he had been "starving to death" and went to work in the oil fields. Lawson worked under his father until he was seventeen, then went to Illinois and from there to California and from there to Mexico. He returned to West Virginia for two years, then went to Oklahoma. He went back to West Virginia and married the girl who had been responsible for his periodic returns. He worked in Kansas two years and went to Ranger, Texas, during the boom of 1918. There he has remained. Typical too is his eventual settling down after his marriage and the birth of his children.[1]

The driller was the aristocrat of oil field labor. He was in charge of the well and on a cable-tool rig had a tool dresser, or a "toolie," as a helper. On a rotary rig he had four or more roughnecks under his command. Although a bright and willing toolie or roughneck might, particularly in boom times, be advanced to drilling after a short apprenticeship, so rapid was the expansion of the oil industry that for more than fifty years competent drillers were in short supply. This bred an independence comparable to that enjoyed by steamboat pilots in Mark Twain's riverboat days. If the driller didn't like what went on he could tell his boss to go to hell, and find a new job. He often did.

Too, drilling was often an entry into the oil business. If a driller had a good reputation for competence and reliability, he could probably get a contract to drill. And if he had a contract at a price that promised a profit, the oil well supply companies would sell him a rig on the credit. He would work one tour himself and hire another driller and crews. If he was successful he would in time buy another rig or two, and devote his time to supervision. Eventually he might acquire some oil properties, either by leasing land and drilling or by taking an interest in a lease as part of his pay for drilling. E. I. Thompson, former executive officer of the Texas Independent Producers and Royalty Owner's Association, believes that 90 per cent of the oilmen came up "the hard way," that is through work in the field.[2]

But whether he was ambitious to become an oil magnate or was content to remain as he was the rest of his working life, the driller realized his importance. The oil industry was still young when reports came from the oil fields about his pride, manifest among other ways in his attention to dress. He might wear denim on the derrick floor, but he was particular about his hat and shoes. When common work shoes were available at $1.75 a pair, he paid $12.00 for fine quality laced boots. He wore a good hat, but it must identify him as a driller. This meant that even when he was off duty his hat must be spattered with slush. "Every other article on his person," wrote a journalist in the *Oil City* (Pennsylvania) *Derrick,* "must be immaculate but his hat must have the white spots from the sand pump or the wearer will not consider himself dressed up." If necessary he would take a new hat to the slush pit and put on a few spots by hand before he would deign to wear it in public places.

He had other extravagances. In the early days men were paid not weekly or monthly, but upon the completion of the well in cash on the derrick floor. "It was the custom," according to the *Derrick,* for the driller "to take his Sunday clothes to the derrick with him on that day, and when the money was received, to strip himself of all clothing he had worn on the well and cast them aside. . . . Boots that cost him $12.00 a pair were thrown aside as carelessly as his cheaper overalls." The custom, the *Derrick* observed, made business for local merchants and "often solved for a less fortunately situated resident the question of winter clothes."[8] In 1942 Joe Moon, who had then lived ninety-one years and had spent a good part of them drilling in Pennsylvania, said the custom was not so extravagant as it sounded. When he brought in a good well, Dan Moran, for whom he worked many years, would give him a new suit. He implied that the custom was not universal.

It did not reach the Southwest, but the recollections of the early drillers afford many examples of the driller's pride and independence, manifest both in the indulgence of personal idiosyncrasies and in a sometimes belligerent assertion of prerogatives.

There was Dad Titus of the Tituses of Titusville, where the first oil well was drilled. In South Texas he worked for Walter Cline. He was called Dad because he was older than the other men. One of his peculiarities was his preference for the graveyard, or night, shift. His experience and seniority entitled him to the day shift, but he refused to take it. He would work the night shift or not at all. Each night, before he went on duty at twelve, he ate a good meal and was given a pail in which the landlady had packed his breakfast, to be eaten at about seven o'clock in the morning. When Dad Titus got to the well, he would put on his work clothes and take his seat on the driller's stool. But before he started drilling he would place his lunch-pail between his knees and look through it. If there was an apple or an orange or banana, or a piece of cake or pie, he would lay it aside, close the pail, and hang it on a nail. Then he would return to the stool and eat whatever he had laid out.

Cline once asked him why he did this. He could not be hungry so soon after the meal at the boardinghouse.

"Well, son," he said, using the usual preface to his philosophizings, "I am a good deal older than you are, and I've been around drilling rigs longer than you have. You'd be surprised how many things can happen to your lunch. The rig can burn down and you won't have anything to eat. The ants can get into it. It can fall off the nail and a stray dog or a hog can get it. There's so many different things that can happen to it that when I get out to the rig, I look through my pail and see what's in it that I know I would like to have if I could save it until breakfast time next morning. But I just can't afford to take the risk so I eat it right then."[4]

And there was Grant Emory, who learned his trade in Pennsylvania, worked in the principal fields of his day, and became a legend before his retirement in the 1930's, and who gave the oil industry one of its classical folk expressions. The two men who operated the cable rig had to have help when casing was set. On one well Emory sent word to his superior that it was time to set casing and asked for a casing crew. The superior waited until school was out at four o'clock and sent over some high-school boys to help run the casing. Emory pulled up the first joint, lowered it into the hole, and set the spider, or clamps, to hold it. Then he turned to the boys and said, "School's out." This expression was taken up by others and was widely used to mean, "Crew dismissed. No more work today."[5]

When retired drillers talk of the past, you may expect accounts of run-ins they have had with the higher brass. Billy Bryant's boss had a brother-in-law who wanted to be a driller. The boss wanted him to work the day shift so he could be with him and teach him. But Bryant's seniority entitled him to the choice of tours. He said, "You didn't stay with me in the daytime. You made me flat-foot it [serve as roughneck] two years before you gave me a drilling job. So you better let your brother-in-law run nights and you run nights." Bryant left then and there, and went to work for another man.[6]

C. C. McClelland was working for a contractor named Tom Wayland, who was drilling for the Blackwell Oil and Gas Company. McClelland had the day tour. Shortly before midnight, when he should have been relieved, he hit gas. The other crew did not show up; McClelland drilled into the sand and pulled out the tools. Early in the morning the company superintendent arrived with ten roustabouts. Gas had been anticipated and a line had been laid to the derrick floor to take it away. When they got everything arranged the superintendent told McClelland to shut down the well. The roustabouts were evidently afraid of the gas. McClelland told him to go to hell. What were those roustabouts for? He wasn't working for him anyway. He was working for Tom Wayland. It could blow as long as it wanted to. He wouldn't shut it off. He went to the boardinghouse and found the crew that should have relieved him sleeping off a drunk. He threw two buckets of cold water on them. In the meantime the men at the well had closed the valve. It didn't take ten minutes to do it.[7]

When the field superintendent of a company Cotton Young was working for hired some men at $2.50 a day to replace men Young had hired at $3.00, there was an argument. It resulted in Young's telling the man that except for his gray hair, he would pull him out of his buggy and mop up the ground with him. The superintendent said he wanted to make it clear that he would do the hiring and the firing.

And I said [Young relates], "Well, by God, this one time you ain't going to do it, old man, not while I stay here. I'm going to have something to say about these men myself."

Of course drillers in them days [Young continues] had the right to hire and fire their own men, and that way they kept a good crew that they wanted and liked. But he wanted to change it all up. So he fired me and I left there.

He was recalled by higher officials in less than a month.

This was not the only time that Young got mad, but sometimes he didn't get mad when he should have. Recalling one experience he seemed to wonder why he didn't. Young was then a drilling contractor. A well had been brought in and it was time to set casing. The company man asked him what he needed. "I'll get you anything you want," he said, "but I want you make it hold." This was equivalent to questioning Young's competence — like asking him if he could drill through quicksand. "Of course it held. There wasn't any doubt about that."[8]

The taciturnity of the driller was legendary. For one thing, it was difficult to talk above the noise of the machinery, and if the men were all experienced and knew their duties, they might work through a tour with hardly a word. Another thing that discouraged talk was the curiosity of the oil scouts in the days when operators kept drilling information secret to avoid booms in leases. If a scout came by a rig and asked, "How deep are you?" the driller would probably say, "As deep as a tree." Hard-boiled Jim received his nickname from the scouts, who found him a hard case. To one scout he replied, "Not putting out a damn thing." The scout, however, did not take this as dismissal. After a long period of silence, he said, "Looks like rain today." Jim replied, "Yes, I know it, but I'm not putting out a damn thing."[9]

But the classic anecdotes of the taciturn driller pertain to Grant Emory and Cotton Young, and neither involves a scout.

Once Emory had a new tool dresser. At the end of the tour they were cleaning up, getting ready to go to town. The tool dresser remarked, "Mr. Emory, we made a pretty good run today, didn't we?" Emory replied, "Young man, you talk too much." Another time when clouds were gathering, the tool dresser, again a new one, said, "Mr. Emory, looks like we might have a storm." Emory's reply was the same, "You talk too much."[10]

One morning Cotton Young and Jess Lincoln left Electra to drive twenty miles to a well. As they were leaving town, Lincoln said, "Cotton, you remember that old bulldog I had?" Though the two men were close friends, Young did not reply, and both remained silent until they reached the well. Then Lincoln said, "He died."[11]

Whether or not the driller deserved the reputation he enjoyed as a drinker and fighter is an open question, for the testimony is conflicting.

The qualifications of an early-day driller [says H. P. Nichols, who drilled his first well at Spindletop in 1902] were determined by the amount of liquor he could wobble around with and the number of fist fights he could win. While not a teetotaler, I was never drunk in my life. Between the ages of eighteen and twenty-five I admit a tendency to become embroiled in a nice clean fist fight. They were for the entertainment of the crowd and usually staged in some sort of joint. My nose has been broken numerous times, and for several years I carried the distinctive mark of a willing but losing gladiator— a pair of black eyes.[12]

A. B. Patterson, referring to the period from 1905 to 1909, says there was more drinking then than now. An employer could hardly enforce temperance upon his employees, for he was glad to get anybody he could to work.[13] Sam Webb, recalling his experiences at Spindletop, says everybody drank, or nearly everybody. Walter Cline describes the early oil field workers as "big, hard and rugged." "They lived pretty hard" and "drank more than you'll find the average boy drinking now."[14] Cotton Young drank freely with one crew he worked with, and once went on the rig hardly fit for duty. He "twisted off," i.e., twisted the drill pipe in two. He was not fired, but his next job was with a crew that never drank, and he didn't either, then or later. After he married he didn't average one drink of liquor or one bottle of beer a year. All he drank was a little eggnog at Christmas time. But if he had stayed long over on Black Bayou in Louisiana, he guessed he would have been a regular liquor hound.[15]

W. J. Rhodes knew a driller whose capacity for liquor was a marvel. He worked the tour from midnight to noon. Each day as soon as he came off duty he got drunk. Before his tour began at midnight his tool dresser would come for him in a buckboard. With the help of the driller's wife, he would load him in the buckboard and drive him to the well. He would work until noon, drilling more hole per twelve-hour tour than any other man in the field. When he came off duty he would get drunk again. He did this daily for a period of sixty-three days. He did not live to a great old age.[16]

But drinking of this sort was rare. Typically the driller who drank got off from work for a periodic spree. Allen Hammill used to employ a driller whom he called Fred. He was a "dandy" driller on the job, and he would stay on the job until he had accumulated a few hundred dollars. Then he would come to Hammill and say that he had to get off for a while; he had to go to town. He would come back pretty well worn out, but ready for work. He would save until he had a few hundred dollars, and then go off on another spree.[17]

Big Hole Bill was a cable-tool driller who boasted that he was a big-hole man. The bigger the better, and he refused to work on any less than ten inches in diameter. When it came time to reduce the hole below this size, he would quit. He would leave little holes to little men. This was his boast, but he did not fool his peers. Every driller knows that the big part, that is the beginning of a well, is easier drilling than the deep part. It is when you get down a thousand feet or more that you need all your skill. The real reason Big Hole Bill would quit was to draw his pay and go on a spree.[18]

Peck Bird worked for the Hammill brothers on the Lucas well. He has been called a good driller and a good man. He did not, however, save his money, get into the oil business, and prosper as his employers did. When he died his friends took up a collection to bury him. When all the expenses of the funeral had been totaled they found that they had sixteen dollars left. "What will we do with this sixteen dollars?" one asked. "I know what Peck would want us to do with it," another said. "He would want us to buy a keg of beer and hold a wake." And that is exactly what they did.[19]

Another driller whose friends took up a collection to bury him was Grant Emory. He became increasingly addicted to liquor, which after his working days were over was hard to come by. How many collections were taken up for his funeral before he died I do not know. There were at least two. Once an oilman was approached by two old drillers known as Yellow Young and Toledo Jack. They said, "Grant died last night. You heard about it?"

"No. Hadn't heard it."

"Yes, they found him dead in bed. We're taking up a collection to bury him. We thought you might want to kick in."

The oilman gave them twenty dollars. Later in the day he saw Yellow Young, Toledo Jack, and Grant coming up the street with their arms around each other singing.

Some months later Yellow and Toledo came to him again.

"Have you heard about Grant?"

"No, I haven't heard."

"They found him dead in bed this morning. We're taking up a collection to bury him. We thought maybe you'd like to chip in."

"The first time Grant died, I gave you twenty dollars. This time I'll only give you ten."[20]

No doubt the drillers furnished their quota of customers of

the saloons and speakeasies that often outnumbered the grocery
stores in the boom towns. They were seldom habitual alcoholics.
If they reached this stage, they became ex-drillers. And, clearly,
one did not have to drink at all to maintain status.

Nor did he have to be a fighter, though tales of fights will be
told when old-time drillers get together.

One source of conflict, generally good-natured, but sometimes
an occasion for a fight, was state pride. At first the Pennsylvanians regarded themselves as the elite. They referred to the
West Virginians as snake-hunters, afterwards shortened to snakes,
and said they knew nothing about the oil business and were too
dumb to learn. The West Virginians called the Pennsylvanians
starving owls. They were starving in their own state and had to
come to a better one to make a living. They adopted their own
nickname, and one of them, a man named Rogers, made up a
song with this refrain:

> We're the boys that fear no danger.
> We're the West Virginia snakes.[21]

When oil was found in Illinois, the Pennsylvanians and the West
Virginians joined together in calling the workers from that state
scissorbills.

Something of the Yankee-Rebel antagonism survived. In North
Texas one hill upon which a Yankee crew was drilling was called
Bluebelly Hill. Native Texan Cotton Young wanted it known
that he didn't have anything against the Yankees—some of the
best friends he had in this world were Yankees, but some of them
when they first came down from the East didn't have any sense.
They thought they knew everything and the fellows down here
didn't know anything.[22]

Only partly sectional was the rivalry between the cable-tool
men and the rotary men. The cable tools, developed in West Virginia by the salt well drillers, were used on the Drake well and

all subsequent wells in the North. The tools, consisting essentially
of a rope socket, a drill stem, and a bit, are suspended on a cable
and are raised and dropped in the well. The rotary works on the
principle of an auger. This equipment had been developed in
the 1890's at Corsicana, Texas, where production was not large
enough to attract great numbers of workers from other states.
It was used successfully at Spindletop, after cable tools had failed
on account of quicksand, and in other coastal fields. Its area was
greatly extended with the perfection of the Hughes rock bit in
1909. It did not, however, completely supplant the cable tool rig,
which is considered superior for drilling certain formations. One
field may be a cable tool field, another a rotary field, and in some
fields both are used. The rotary men called the cable tool men
rope chokers, jar heads, and mail pouchers (from Mail Pouch
chewing tobacco), and to the cable tool men, rotary men were
auger men, chain breakers, clutch stompers, twisters, and swivel
necks. At Electra, Texas, where both types of tools were in use
in 1911 and 1912, some of the boardinghouse keepers resorted
to segregation to maintain a measure of tranquillity. One house
would be for swivel necks only and another would be reserved
for mail pouchers.[23]

Yet these antagonisms among the oil workers were not so
deep as they might appear to an outsider. If there were occasions
for banter and fistfights, they were more often occasions than
prime causes. For there were among the oil workers men who
enjoyed a good fistfight, particularly after a few drinks. This
was one release from the tensions of hard and dangerous work.
H. P. Nichols, already quoted, staged his fights "in some sort of
joint" "for the entertainment of the crowd." "In the old days there
was much fighting," recalls O. G. Lawson. "Men were of a rougher
character. They just seemed to go in for that. Of course, there was
no other interest much of any kind."[24] William Cotton had seen
fellows fight lots of times when they were pretty full, just to see
who was the best man. Sometimes they might put up ten dollars

and sometimes nothing. They weren't mad when the fight began and they still weren't mad when it was all over. They just fought for fun.[25]

The most famous of the fighting drillers was a Bill Quinlan, who worked for years in the eastern fields. He fought for fun, and his fun was the greater because he won every fight he started. Then he moved to Coalinga, California, where his next-door neighbor was a preacher. Their children got into a fight, and the fathers got into a fight over whose children were to blame. After a long indecisive struggle, Quinlan said, "I believe you are the toughest man I ever fought. I don't know whether I can whip you or not." The preacher said, "That is just what I was thinking about you." They fought a while longer and called it a draw and became friends.[26]

The driller, in short, went through a stage similar to one that the trapper, the miner, and the cowboy went through, one in which a chief satisfaction was a pride in his manliness. He was changed by changes in the requirements of his trade.

A few years ago I was talking to a driller who had retired several years earlier after forty years' service. He pointed to a house on the next block and said, "A driller lives in that house. He is thirty years old and he has already drilled more wells than I did in my forty years." He said he would be at a complete loss if called upon to operate the tools the young driller uses.

The revolution in equipment, along with the rise in the general educational level, has changed the character of the oil field worker. Physical size and strength are no longer important. The screwing and unscrewing of the drill pipe and all lifting are done by motors. A man weighing 130 pounds can do as well as one weighing 180 pounds. The guesswork—which might also be called the art of drilling—has been largely removed. For example, in rotary drilling the full weight of the drill pipe must not rest on the bottom of the well. If it does it is likely to break, and in any event, the well will be crooked. The old driller estimated the weight by sound, by the speed of the rotary, and by the behavior of the pipe. The present-day driller reads the weight on a dial. Drillers and roughnecks are now skilled technicians. They take pride in their work, probably as much as their predecessors did, but they are less bumptious in their manifestations of it. Yet the old folk image persists, and the new driller is hardly known outside the oil industry.

12

The Landowner

UNTIL RECENTLY folklore made the landowner an unsophisticated, highly suspicious countryman, who when oil came passed overnight from rags to riches. Riches attained, he and his family had difficulty in adjusting to their new economic status. They became comic figures either because of the persistence of old and now inappropriate habits and attitudes, or because of their too ardent pursuit of new ones. Their wants remained ludicrously modest in relation to their means, or they spent their money in riotous living.

It was the oilmen who were chiefly responsible for the tradition of suspicion, for it was the suspicious landowner who created the incidents most vividly remembered.

Long before it reached Texas and the other southwestern states, the industry, for reasons not always valid, had come to be regarded with caution. Standard lease forms were slow in developing, and in the absence of such forms the landowner did not always know what his rights were. Tales of fraudulent companies had circulated throughout the country.

Illustrative of this distrust is the widely diffused legend of the plugged well. For years wherever unsuccessful wildcats were drilled rumors would spread that oil had been found but that the hole had been plugged or the oil cased off by the driller, who had been bribed, usually by an agent of Standard Oil, but which of the operating companies he represented was never divulged. Sometimes the rumor would be that the well had been plugged, either because the company had sufficient oil to meet demands and did not want new production or because it wished to lease additional acreage at a small rental. In that case you could expect it to return in five or ten or twenty years and develop the field.

A. E. Ungren, an independent producer who was a driller before World War I, drilled a well in Kentucky just before going into the military service. The well was as dry as any ever drilled. The last formation he went through was a black shale, and the slush had something of the color of crude oil. When he returned from the war and visited the community, or rather a girl who lived there, several people told him about the well, how oil had been struck and the son-of-a-bitch driller had cased it off. They did not know that Ungren was the driller and he did not tell them.[1]

Probing in what became the Premont field in Texas went on for more than two years before commercial production was obtained; and rumors began to be heard among the ranchmen that there was plenty of oil and that the company was deliberately concealing it, that the wells were being plugged. They were more convinced than ever when, on October 29, 1934, a well that was being reworked blew in, cratered, and caught fire. One ranchman was quoted as saying, "But they let this one get away from them. There wasn't nothing they could do to hide it. All they could do was try to put it out."[2]

But the prize skeptic was a lessor in Louisiana. Determined to protect himself against any possible skullduggery, he demanded that a clause be put in the lease requiring the company to furnish him an electrical log of the well. The accommodating lease man, deciding to go the lessor one better, persuaded him to accept the following agreement instead:

If a well should be located on the surface of the land herein leased, then in lieu of the Lessor being furnished an electrical log, Lessor shall have the right to be lowered head-first down the casing equipped with a two-cell battery flashlight, to a depth of 7,000 feet, or some lesser depth if heaving shales be encountered. Lessee agrees that Lessor will be lowered at a rate of speed not to exceed that rate which any prudent operator would lower his Lessor into a well bore. It is further understood that if lowering line should part, immediate fishing operations shall be commenced, and in no event shall Lessor be cemented in plug and abandoned.[3]

I have asked dozens of oilmen whether they have known of the plugging of a potentially producing well. The almost unanimous reply is that they do not believe such a thing ever happened. Why would one drill in the first place if he did not want oil? Two informants, however, cited instances. In both cases the well had been financed at considerable profit to the promoter by the sale of fractional interests in the well. The fractions sold, when added together, totaled more than 100 per cent. If the wells had been allowed to produce, the promoters would not only have had liabilities to meet beyond their means, but would in all probability have been sent to the penitentiary. So they plugged the wells and left the country.

In this atmosphere of suspicion the task of the lease man was sometimes hard. Walter Sharp paid in currency. Ed Prather always carried cashier's checks. He recalls a trip to North Texas on which he carried $112,000 in such checks. At one strategic place he spilled the checks on the floor as if by accident as evidence of his intent and ability to pay. When he went to a community he would call on the banker and secure his co-operation. He would assure the landowner that he was acquiring the leases not for speculative purposes, but with the intent to drill, and as evidence of his good faith he would leave the executed leases in the custody of the banker.[4]

When Joseph Weaver began operations in Texas, associated with him in the leasing of land was a native Texan, Jim Leach of Fort Worth. Early in their search of acreage they called on Mr. and Mrs. G. W. Fisher, who owned a quarter-section of land in the Eastland-Ranger area. Weaver, intent upon gaining their confidence, launched into an eloquent speech. He told them that if they signed the lease they would not be looked upon as outsiders, but as associates whose interests were identical with those of the company, that the lease was being sought for purposes of development, not for sale. While Weaver was delivering his speech, Leach was sitting silent in the corner of the room. Fearing that Weaver had spread it on so thick that his sincerity might be questioned, Leach spoke up. He said, "Mr. and Mrs. Fisher, I've never seen you before, but if you throw in with that crook that's been talking to you, you're throwing in with the worst crook outside of hell."

Weaver thought he was ruined, but Fisher looked up and said, "Well, if that's the case, I'll sign the lease."[5]

Another experience of Weaver's concerns a lessor upon whose land had been drilled several wells which in a short time became very small producers. The landowner blamed Weaver for the decline. Once when he was in the East, Weaver's secretary wired him that the man threatened his life. When Weaver returned he immediately had a visit with the man and soon all their differences were settled. Weaver then asked if it was true that he had threatened to kill him. The reply was, "Joe, I may have done that but if I had killed you I would have known that I killed a good man."[6]

There was a man named Knight who owned a small farm, possibly twenty acres, on Batson's Prairie. He could not read and signed documents with an X (his mark). He lived with his wife and four or five children in an unpainted shack. When oil was discovered at Batson, various oil companies tried to lease his land. He flatly refused to consider a lease. If they wanted to drill on his land they would have to buy it. The price he set was between forty and fifty thousand dollars. Finally one company agreed to pay his price. Knight would not take the check offered him. He told the lease man to bring the money in currency so he could count it and see that it was all there. The man went to Beaumont to get the money, and arrived at Knight's house with it late in the forenoon. Mrs. Knight was cooking dinner. Knight drove in from the field and left the team hitched to the plow. They sat down at the kitchen table and counted out the money. Knight was satisfied, his wife signed the deed, and he made his mark. He immediately went out and unhitched the team from the plow and hitched it to the wagon. Without packing a bag or taking any clothing except what they had on, he and Mrs. Knight and the children got in the wagon and headed toward Liberty, twenty miles away. They did not wait to eat. The pots and pans were left boiling on the stove.

They got to Liberty in the afternoon and hitched the team to a rack near the railroad station. Two or three trains going east and west passed through Liberty each day. Nobody remembered seeing the Knights get on a train, and the ticket agent could not remember selling them tickets, but the Batson boom was on and he did not have time to take note of all ticket buyers. At the end of the day the team was at the hitching rack, but the Knights were gone. They were never heard of at Batson or Liberty again.

And so the mysteries remain. Why did the Knights refuse to lease their land? Why did they demand cash? Was it because they distrusted oilmen? Why did they leave their personal belongings behind? Why didn't they sell the team? Why did they leave so hurriedly? Was it because they hated the farm that had supported them so poorly? And what became of them? Did they go somewhere and find a farm big enough to support them in comfort? Did they meet foul play? Efforts to trace them failed.[7]

There is a vast body of legend on the rags-to-riches theme. The first fallacious assumption in this legend is that anybody who has an oil well on his land is rich. Obviously the income will depend upon (1) the daily production, (2) the life of the well, and (3) the price of crude oil. In 1902 crude oil sold in Beaumont for as little as three cents a barrel; in 1917, with the war on, crude brought three dollars a barrel; but in 1931 in East Texas, it dropped to ten cents. The landowner, under the standard lease almost invariably used, gets $12\frac{1}{2}$ per cent of the oil. If his land is favorably located with respect to a producing well, he will receive a high rental fee, or a cash bonus. Thus he might be paid a thousand or more dollars per acre for the right to drill, but this payment ceases when a well has been completed, and when rentals of this magnitude are involved, drilling will not be delayed long. Let us say that one had a fifty-barrel well on his land which before proration laws were enacted by the oil producing states was permitted to flow at full capacity. His share at $1.25 per barrel would be $7.81 a day, an income not to be scorned, but somewhat less than the wages of a top driller. At 1960 prices and with the average daily production of fifteen barrels that prevails in Texas, the daily royalty amounted to five or six dollars per day per well.[8]

The well might produce a year or two or three, or even twenty with some decline in production. One well was brought in at an initial production of 5,700 barrels a day. It declined steadily and in a year ceased to flow. It was placed on the pump and in six months the production was zero.[9] Again assuming $1.25 oil, the owner's income began at $890.00 a day and reached zero in eighteen months. This would not have happened under now existing proration rules. For August, 1958, the 184,507 producing wells in Texas were allowed to produce 2,936,026 barrels of oil, an average daily production of 15.9 barrels per well.[10]

G. W. Taylor came from Oklahoma to Eastland County, Texas, in 1907. He bought a quarter-section of hillside land upon which he hoped to make a living farming. Anybody who knew the country could have told him that he could hope for little more. He built a barn which was to serve as a residence until he could build a good house. But the years went by, his children were born and grew up and left home, his wife died, and he was still living in the barn, which now leaned acutely and was propped up with heavy timbers. He had passed seventy when oil was found on his land. The production was small. He has received monthly checks for as much as $700, but in 1956 they were less than $400. He was then living by himself as he had for years. As for building the house, there was no point in it now. His children wanted him to take a hotel room or apartment in Cisco, eight miles away, but he would not feel at home there. He could afford some kind of car, but had never learned to drive and had no desire to learn at eighty. Besides, his neighbors and the motorists on the highway were very good to him. He hitchhiked to town about twice a week to buy groceries. In summer he spent most of his time sitting in a rocking chair outdoors on the shady side of the house reading novels. His eyesight was good and he was not interested in radio or television.[11]

It is true that oil has been found on the land of owners whose poverty would be hard to overstate; but just as folklore exaggerates the wealth of the landowner after his well comes in, so it exaggerates his poverty before. Examples will come to the mind of anyone who is familiar with the lore of an oil community. There is Mrs. K., who would not wish to be mentioned by name. After she had been made wealthy by wells on her ranch, she was called an ex-truck-driver who got rich in oil, but how she managed it was not specified. Now nearly any woman who has lived on a family-operated farm or ranch has driven a pickup on occasion, and I have known some who drove heavier trucks during harvest time when labor was scarce. But not many women have made a living driving a truck. Certainly Mrs. K. did not.

And there is Mrs. McCleskey. Soon after the McCleskey well came in, a Fort Worth newspaperwoman called on her in Ranger. Among other words that she put in Mrs. McCleskey's mouth were these: "I got up early every morning. We both worked hard. I made quite a little money taking in washing."[12] I have interviewed a number of Mrs. McCleskey's neighbors, and they are unanimously emphatic in saying that while she did her own laundry, she never did washing for others.

There is similar inaccurate reporting in Olive McClintic Johnson's article on the Mexia boom in the *New York Times:*

There is Professor Reid, the village pedagogue, whose stock in trade was spelling, and yet who never in his wildest flights of fancy thought that the 117-acre farm he managed to buy on the Tehuacana road would spell riches for him. Up to last year about all the farm ever spelled for Professor Reid—what will [*sic*] renters, taxes, droughts and boll weevils—was trouble. Now the professor threatens to be obliged to change his specialty from orthography to mathematics in order to keep track of his royalty checks.[13]

Reid at some time in his life might have taught in the village school, but if so, it was a long time before oil was found on his farm. At that time he was professor of chemistry at the West Texas State Teachers' College, and before his appointment to this post he had held an executive position in the State Department of Education. Neither of these positions paid him a salary that would place him in the ranks of the economic royalist, but his income was somewhat more than that of a village teacher. Furthermore, he had had flights of fancy. His farm had not proved to be a very remunerative piece of rental property, but he had kept it in the hope that oil might come sometime.[14]

Even though oil does not always mean great wealth, the old metaphor dating back to the time of the legendary Dick Whittington and his cat has in the oil regions been largely supplanted by another of the same import. One does not say, "When my ship comes in," but, "When my well comes in." "When my well comes in, I'm going to have a new hat." Or it may be a car or a new house or a trip to Europe. Everybody is entitled to use this expression, including those who own neither land nor royalties nor leases, nor even stock in an oil company, and who have no more prospect of having a well come in than they have of a ship coming. But a lucky few do have prospects and a lucky few do have wells come in. What do they do when their wells come in? Or even before their wells come in, if they have received a considerable rental or bonus?

An East Texas farmer is reported to have taken his first check to the bank and gone shopping. He made two purchases: a forty-dollar Stetson hat and five dollars' worth of bananas.[15] The story may not be true, but it is plausible, and in its essentials it has happened many times wherever oil has been discovered. In the culture in which he lived the hat was a symbol of status; the bananas a luxury hitherto enjoyed only on special occasions, like Christmas and Thanksgiving. The symbol of status may be an expensive car, membership in an exclusive club, or a residence in River Oaks or Highland Park. The luxury long longed for may be anything from a new cookstove to a trip to Europe. These things are attended to first.

The East Texan was not the only man to buy bananas. A few days after the well came in on John McCleskey's farm at Ranger, he went into Gordon's store and asked for a stalk of bananas. Asked why he wanted so many, he said that one time in his life he was going to eat all the bananas he wanted. He took the stalk to the sidewalk in front of the store, placed it on the bench where the whittlers sat, and called all the children in hearing distance and told them to help themselves to all they could eat. He himself ate five or six.[16]

One farmer was a great coon-hunter and the owner of thirteen dogs. When his well came in, he said that he was first of all going to buy his dogs all the red meat they could eat. Then he was going to get an automobile. His friends would be welcome to ride with him if they could find room after his thirteen dogs got in the car.[17]

Two sisters bought nothing for several months after their wells came in on the farm that had barely supported them. Then one day they appeared at the bank saying they wished to draw out their money. The teller, knowing the size of their account, referred them to the highest official on duty. The banker did not wish to lose their account, and besides he was hardly prepared without previous notice to hand out currency in the amount of their balance. Furthermore he was afraid that rumors might start a run on the

bank. He asked the sisters into his private office. They said they
had full confidence in the bank and were completely satisfied with
the service. It was merely that they had decided to go shopping in
the city and would need their money. They said they had a considerable list of things to buy and they didn't know how much it
would all add up to. The banker estimated prices and totaled the
list. It came to a few hundred dollars. The largest item was
linoleum for the floor.[18]

A blind Negro used to beg on a street corner in Mexia. When
a well came in on his farm he bought two cars, a Cadillac and a
Cleveland. He continued to frequent his street corner, but he no
longer begged. His sons would bring him to town in the morning
and call for him at night. He never learned to distinguish the
cars by their horns. His question each night would be, "Which
car you driving this time, the Cadillac or the Cleveland?"[19]

Cars, bananas, red meat for coon dogs, cookstoves, linoleum—
when the wells come in the landowner buys something he has
long wanted but which has hitherto been beyond reach.

Surprisingly, the new ax has become the most widespread
symbol of the new wealth. The story is meant to be comic, the
humor depending upon the discrepancy between the wants and
the means of the formerly poor but now rich landowner, with an
additional touch of irony in that oil brings a new fuel and makes
the family ax obsolete. I first heard the story in the McCleskey
version from Ranger.

In 1917 Uncle John and Aunt Cordie McCleskey, as their
neighbors called them, were living on a farm a mile south of
Ranger. They had an unpretentious but substantial farmhouse
with a garden and orchard, for like most of their neighbors they
produced much of their food at home. For cash income John
McCleskey grew cotton and peanuts, and worked as a bricklayer
on the rare occasions when there were any brick to be laid in the
little village. He was not used to handling large sums of money,
but his condition was hardly one of poverty. His property was

valued at some $20,000, which in 1917 placed him among the more prosperous farmers of his community.

It was upon his farm that the Ranger discovery well came in on October 22, 1917, with a flush production which gave him an income of about $250 a day. Some weeks after his good fortune he built a cottage in Ranger, where the McCleskey Hotel, representing the first sizable investment of his oil money, was under construction.

It was at this time that a newspaperwoman from Fort Worth called upon Mrs. McCleskey for an interview. The reporter quoted her as saying that when the well came in, her husband asked what he might buy for her. Mrs. McCleskey replied, "Well, the blade of the old ax has a nick in it and I would like to have a new one to chop kindling with."[20]

I have heard the story in several variants (in one the handle of the ax is so old and rough that it leaves splinters in her hands) from at least a dozen informants who believe it to be true. I accepted it as truth myself until a few years ago when on the trail of Gib Morgan I found the same story in Pennsylvania. I have since heard it from Desdemona, where a farmer is reputed to have said that he was going to get not only the best ax that money could buy, but also a grindstone and a gasoline engine to turn it. Then he was going into a thicket and see how it felt to attack a post oak with a really sharp ax.[21] Bob Duncan has a version of the ax story from Beaumont.[22] Haldeen Braddy heard it in Vann. George Sessions Perry must have heard it, for in one of his short stories he has a character say of another who had oil land in East Texas, "He's got so much money he can't hardly spend it. He bought three new axes an' a barrel of lamp oil an' a barrel of flour, that didn't even make a dent in his money."[23]

Doubts therefore arise as to whether or not Mrs. McCleskey did in fact ask her husband for a new ax. She and her husband are dead and their children have left the state. Among their neighbors the story is widely believed, but I have found no informant who will vouch for its authenticity. Mrs. Hagarman, the widow of the first mayor of Ranger, for example, said that since Mrs. McCleskey did not consider herself too good to go to the woodpile and cut up an armful of wood, the story was not impossible. She said, however, that many stories were told about the McCleskeys which she knew to be false. She said that at one time in the Ranger-Breckenridge area "a McCleskey" was used as a common noun to designate any comic story of the newly rich.[24]

John Rust, who lived on the farm adjoining the McCleskeys and was twelve when the well came in, said in a taped interview:

> Yes, I heard the story and I've always doubted that it's true. It could be true certainly, and I'm not sitting here to say that it's not a true story....I will tell you that I didn't hear her make the remark, but then of course she could have made it...without my presence. But knowing her...and her husband as I have known them for years, all my life, and the kind of people they are...I just don't believe that she made the remark at all. She might have made it, however, in a joking sort of way, just for fun. But to seriously make the remark, I doubt if she would make it.
>
> I know that they didn't do silly things with their money; they were good, thrifty people. They always had plenty to eat and they looked presentable in their clothes. Old John McCleskey was the kind of man that didn't have his ax blade gapped up anyhow. He had an old-fashioned grindstone—I remember where it used to sit—and I've been over there before the oil boom at their home, and maybe old John would be gone somewhere or out in the field working or something, and Aunt Cordie would need some wood to cook the evening meal. I've gone out there and picked up that ax, and I remember that it always had a sharp blade.
>
> And I just have every reason to kind of doubt that she made the remark.[25]

Joe Weaver, retired oil operator who came to Texas from West Virginia by way of Oklahoma, and who was beginning operations in the Ranger area before the McCleskey well came in, said that he did not believe the story. Then he said he would tell me what he called "an old West Virginia story." Oil was discovered on a widow's farm, "and her boys went to Harpersburg to celebrate. They had had a tough time and the widow had been a very hard worker. And the boys were enjoying themselves when one turned to the other saying, 'We must take mother a gift,' and the other one said, 'Well, what in the world will we take her?' And the first suggested, 'I know, we will take her a new ax.' "[26]

It is possible that the story never happened either in Pennsylvania or West Virginia or Texas, but it is a part of the folklore of the oil country.

A woman and her nephew have long been symbols, one of modesty of wants, the other of extravagance. Mrs. Sarah McClintock died in March, 1864, leaving about $200,000 in gold and currency locked up at home in an iron safe and about $100,000 on deposit in the bank, and a farm upon which oil royalties were yielding between two and three thousand dollars a day. (Oil was then selling for more than nine dollars a barrel.) About the only article she seems to have bought after her wells came in was a new cookstove, in which she was starting a fire with crude oil when she received the burn that resulted in her death.

She was a widow without children and her fortune went to her nephew, John Washington Steele, then a few months short of twenty-one, who had been making a good wage teaming in the oil field. In less than a year Steele had parted with his inheritance and was selling tickets for a traveling minstrel show for his board and keep. He had signed various documents while under the influence of brandy. He had played angel to the minstrel show that afterward maintained him. He had rented a hotel and entertained all comers. He had given away horses and carriages. He had walked down the city streets with ten-dollar bills stuck in the buttonholes of his coat, enjoying having them snatched by any passerby who could grab one. These are only some of the means by which he got rid of his fortune.

In a year or so after it was gone, he got a job, became a steady worker, paid out a home, and in 1920 died a sober citizen with a savings account.[27] He is remembered, however, not for his years of dependable service to his employers, but for his months of sensational spending. Steele was an exceptional man, but his pattern of behavior has been repeated with sufficient frequency to establish his nickname, Coal-Oil Johnny, as a common noun in the American language. A Coal-Oil Johnny is anyone who squanders an oil fortune, large or small.

Generosity is usually a noticeable trait in his character. There was, for example, the young West Texan whose friend's wife needed hospitalization which her husband could not pay for. "Take her to the hospital. I have arranged it," said Johnny. "Here is the key to my car. You needn't bring it back. The car is yours." He did not, however, give all his money to his friends. A lot of it went for liquor and cars. They did not mix very well and he never had a car repaired.[28]

One family received a thirty-thousand-dollar bonus when the lease was signed. They left immediately for a trip East. They had a good time staying at expensive hotels, buying at exclusive shops, and seeing the sights generally. In a few weeks their money was all gone and they started home in their new finery, expecting to arrive about the time their first well came in. But it did not come in, nor did further drilling discover any oil on their land. They went back to plowing and planting and chicken-tending with whatever consolation they could find in the reflection that it was fun while it lasted.[29] One woman upon whose land oil was found spent her money as fast as it came in. She bought jewels and furs. She had her old farmhouse torn down and built in its place a Hollywood-inspired mansion, complete with swimming pool. Production declined and then stopped. When the wells quit flowing, the swimming pool went dry. She couldn't pay her water bill. In less than ten years after her first well came in, she was on the relief rolls. She later lived on an old-age pension. She couldn't sell her house or even rent rooms to much advantage. It was out where nobody wanted to live.[30]

A man and his wife had for many years done practically all the work on their farm. The woman would get up before day and cook breakfast. Then she would wash dishes and clean up the house. And then, when crops needed attention, she would go to the field and plow or hoe with her husband until it was time to come to the house and cook dinner. Then she would go back to the field in the afternoon and work until time to get supper. Oil came. A redhead came along and ingratiated herself with the husband, who persuaded his wife "to set him free." The redhead married him but left him as soon as his money was spent.[31] A parallel episode occurred in another field many miles away. Two men began to show attention to the daughter of a farm couple. They included the mother in their invitations to dine out and dance. They took them on motor rides to nightclubs in the city and showed them a world they had never before known. One of the men became engaged to the daughter. The other courted the mother and persuaded her to divorce her husband. There was a division of property. The confidence men got most of the woman's share, but there was no marriage. Later her ex-husband gave her a house and an allowance of a hundred dollars a month.[32]

A man who had lived in a log cabin so nearly ready to fall down that it had props around it, bought among other things a Cadillac and a $9,700 diamond. In about two weeks the well began to fail. A shot of nitroglycerin brought salt water. The big car sat by the log cabin.[33]

"There are always a few who can't take it," is a rather common comment on people of this type. Unfortunately they and the other Coal-Oil Johnnies receive a publicity that creates a false impression of both the character of the landowner and the size of his income.

Another misconception is that a poor man who once owned productive oil land has invariably come by his poverty through extravagance. As a matter of fact, when individual case histories are examined it is found that most of the landowners who have lost their oil money have done so through bad investments, and that not a few have been victims of outright fraud.

In one small town in 1941 an elderly man was making his living doing odd jobs for his neighbors. He once had a three-hundred-acre farm upon which he raised the greater part of his family's food, and sold a few hundred dollars' worth of peanuts each year to provide cash for their modest store purchases. Oil was discovered in his community in 1918 and eight or ten wells were brought in on his farm. Oil was high and the wells were kept wide open. His income for one day was more than it had ever before been for a year. He was bewildered. One day he was seen with his back against the wall of a lumber shed, a half dozen "brokers" forming an arc around him, all of them simultaneously offering him safe and profitable investments. He invested. In 1941 his working capital consisted of a mule and a plow with which he worked his neighbors' gardens.[34]

A contemporary of his was a Negro known in his community as Uncle Zeke. When oil was discovered on his farm, he bought each of his eight children a car, and to each of four daughters he gave a piano. But it was not extravagance that broke Uncle Zeke. A Negro from "up Nawth" appeared in the community and became Uncle Zeke's good friend and confidant. It was not long before he was managing Uncle Zeke's affairs. If he asked Uncle Zeke to sign a document, Uncle Zeke would make his mark. It was not until his bank had informed him that his account was overdrawn that Uncle Zeke learned the nature of the documents he had been signing.[35] Another Negro was brought a paper to sign. The white man who brought it said it was a lease. The Negro asked a white friend to read it. It was a deed conveying the land in fee simple. A lawyer was employed and this man avoided the trap into which many men ignorant of business and shy of lawyers fell.[36] During the depression there were on relief rolls names of men who sold their royalties without understanding what they were doing; and instances have come to light of the guardians of incompetent Indians becoming rich as their wards became poor.

But these cases, always exceptional, have become increasingly rare as the laws have been tightened and several highly publicized convictions have been obtained. No statistics are available, but spot-checking in various oil-producing communities corroborates the testimony of qualified observers that the landowners as a group have handled their oil money quite well.

This conclusion is in accord with the comments of an Oklahoma editor, who in 1924 protested against the inaccurate reporting of the journalists. He wrote:

Among other things that oil has brought us is a flock of newspaper correspondents looking for "human interest" stories about our newly made millionaires. They seem to think that those who have been made rich by the discovery of oil on their land should do something for the entertainment of the readers of the metropolitan press, that they should make monkeys of themselves to make good "copy" for the space writers. The truth of the matter is that Tonkawa's millionaires are behaving very sensibly, and are not making any great splurge with their newly acquired wealth. Most of them are looking after their immediate relatives who have not been so fortunate as they. Some are showing their faith in the soil of this section by investing in farms. They are building business houses, sharing in the stock of hotels and other business enterprises. Not one has done anything that any sensible and well-balanced man or woman would not do. Very few have left the community, but are spending their money here where it was created. They have been besieged with begging letters and even offers of marriage that have been disgusting in the extreme. We want to say right here and now that Tonkawa's millionaires are safe and sane.[37]

Among hundreds of others one could cite Spencer Clemmons, a Negro who in the 1920's attained a brief fame through praise given him in Arthur Brisbane's column "Today." Brisbane got most of his facts wrong and praised him for the wrong reasons. He thought that oil had been discovered on Clemmons' 292-acre farm and he praised him for continuing to work after he had become wealthy. The facts are that oil was not discovered on Clemmons' farm, and he did not become rich. He received a bonus of several thousand dollars from an oil company which drilled three salt-water wells on his farm before abandoning the lease. Clemmons bought a moderately priced car and made a down payment on a house in town. But he used his bonus money principally to send his children to college. Two of them became teachers, one a civil servant employed by the Post Office Department, and two farmers. Clemmons, when he became too old to work, drew old-age assistance.[38]

Another Negro had for years subsisted, but little more, on a ten-acre farm in East Texas. When oil came in 1930, it was discovered that his land sat squarely on the reservoir. He used a part of his royalty to buy another farm, this one forty acres near Hawkins. In 1940 the Hawkins field was discovered on this farm. He deposited his money in the local bank until he was advised that his account was getting too large for a country bank to handle.[39]

Hundreds of individuals could be listed who, according to generally accepted standards, have invested their oil money wisely: in the education of their children, in paying their debts, in increasing their acreage, in improving their farms. An Oklahoma banker says that when people strike oil, they first pay off their mortgages; then they buy homes. Their next purchase is likely to be a new car. Then most of them buy land. A few, but not many, buy stocks and bonds. Only a very few, he says, spend foolishly. County editors in widely separated oil regions say that most landowners who get oil buy farms.[40] A retired oilman, after thirty-five years' activity in all parts of the United States, declares that the men who have shown the best judgment in the handling of their oil money are the bankers and the landowners. The banker invests conservatively in stocks and bonds. The farmer or ranchman, he says, pays his debts, builds a house, and buys more land.

Travelers in the oil regions, seeing farms lying fallow, often assume that the owners have quit work to live on their royalties, and they complain of a moral softening, which they attribute to easy money. Inquiry generally discloses that their assumptions are wrong. Much of America's oil has been found on relatively unproductive agricultural land. And besides, profitable farming can hardly be carried on when a field is under development. So when his oil wells come in, the typical farmer buys land somewhere else. He may return to his native state and county and buy the rich farm that he has long aspired to own. East Texans have left in great numbers, particularly for two regions. Many have gone to the south plains, where with the mechanical equipment oil has made it possible for them to buy, they can grow a bale of cotton for half the cost required on the farms they are abandoning. Many others have gone to the citrus fruit regions and bought orchards. When an oil field is exhausted some of the land is put back into cultivation and some of it is converted to pasture, often its most economical use. Any decline in agricultural production that oil has caused is purely local. Indeed, by supplying capital for better livestock and equipment, oil has enabled many a farmer to produce more than ever before.

The rancher is perhaps even less affected by the discovery of oil on his land than the farmer. Often his liabilities approach his assets, but he is accustomed to thinking in terms of thousands of dollars and is not so easily overwhelmed by his royalties. He pays his debts, builds a new house, and improves his herds. Considerable prestige attaches to his vocation; his way of life is satisfactory to him; and he has no incentive to appear other than he is.

John Steele, the original Coal-Oil Johnny, died in 1920. His successors have virtually disappeared. One reason is that the nature of oil deposits has become widely understood. A few years ago many people believed, science to the contrary notwithstanding, that oil renewed itself. Now everybody knows that no well produces forever. Royalties are not thought of as current income to be spent, but as capital gain to be conserved. Another reason is proration. Under the old system of free enterprise, no rules governed the spacing of wells, and they were allowed to flow wide open. The landowner got his money over a short period, sometimes only a few months or even weeks. A farm that would produce two thousand a month in royalties under the old conditions will now produce not more than two hundred a month. The ultimate royalties will be greater both because the ultimate recovery in barrels will be greater and because the price of crude will be higher. But more important from the landowner's point of view is the fact that he has time to plan for wise investment.

III

Story and Song

IF FOLK BELIEFS *about the finding of oil survive and develop because they have a supposed economic value, and if stereotypes result from oversimplification as a means of understanding, other forms of oil folklore are more nearly related to aesthetic needs, representing as they do an imaginative or fanciful response to the events and conditions of the industry. Tall tales, songs, and unlettered attempts at poetry are obvious examples. Anecdotes are in origin two kinds: accounts of actual happenings that have passed into the oral tradition (e.g., "The Bulldog") and fanciful creations making specific points (e.g., "The Boom").*

13

Song and Verse

AT A CONVENTION held in Beaumont in 1951 to celebrate the fiftieth anniversary of the Spindletop discovery, a group of veteran oilmen were talking of old times. No one present could recall ever having heard a "real old-time oil field song." There was some singing in the oil fields and attempts were made to compose songs about oil field people and their work, but none of the efforts was of sufficient appeal to attain more than brief and local distribution.

One oilman attributes this sterility to the decline of singing in general:

As a boy you'd pass the fields and the cotton choppers would be singing a song, or when they picked cotton, they'd sing a song, or if a bunch of men were working on the railroad, they would sing, or cutting wood, they sang, or your cook, she would sing in the kitchen. But you don't hear that; nobody sings any more.[1]

He thinks the phonograph and the radio are in part responsible, yet the oil industry had existed for more than a half-century before either of these instruments was generally available.

In the oil industry the conditions that favor the creation of folksongs were absent. First, there was lacking a sense of community. The workers came from various environments, and their associations were temporary. Crews formed and dissolved. Each man knew his place in the crew, but shared no years of common tradition with his fellows. Then too, most of the traditional themes of ballad and folksong were irrelevant. There were, indeed, sensational accidents resulting in loss of life, which at an earlier time and in a different culture (Texas-Mexican and Southern Mountain,

155

for examples) would have produced broadsides. But the broadside was not a news medium in the oil fields, and death was too common to elicit terror or wonder. There was nothing bucolic in the boom towns and oil camps to produce a "Home on the Range"; and the lawlessness was hard to fit into the pattern of Robin Hood and Jesse James. Operating a drilling rig lacked the romantic appeal of work done on horseback. Each man had his job, and there was little work in which singing would be useful in establishing a rhythm, as in chopping cotton, rolling bales up a gangplank, or hoisting a sail.

The exceptions would be setting casing or drawing or returning drill pipe in the days when joints were screwed together by hand tongs. Hardeman Roberts is one of the few drilling contractors who reports having heard men sing at these tasks. While drilling in India, he employed native laborers, who in general were smaller and less strong than the American roughnecks. When they would set casing, they "would take a heavy chain and put that around the casing . . . two or three times, then tie it . . . around a big pole with five or six men on each end." Roberts hired a man whose duty it was to lead the singing as the men in rhythm pushed on the pole to screw the casing together.[2]

But when drilling was going on, whether with a cable tool or rotary rig, there was too much noise to encourage conversation, much less singing. The quiet solitude that encouraged the cowboy to make songs was lacking.

Nevertheless there were opportunities to sing and many workers did sing. They sang "whatever happened to come along,"[3] including old hymns and the late music hall hits. "The Wearing of the Green" was popular, as were "The Bollweevil Song," especially timely at the beginning of the century, and "Frankie and Johnny," which Early C. Deane heard at Batson in 1903, and which he thought at the time pertained to a local woman who had killed her husband under circumstances similar to those detailed in the song.[4] About the same time Max Schlicher got together with three or four other young men and they formed "a kind of quartet."

"We'd go through town," he recalls, "and sing our quartet going home. Then we'd stop close to home under some old pine tree and sit there two or three hours till twelve or two o'clock in the morning, singing old songs together." Their favorites were "Sweet Adeline" and "Harvest Moon."[5]

Attempts were made, however, to create oil field songs, and some of these attempts have a folk quality which under more propitious circumstances might have given them wide oral circulation. Among these are Hardeman Roberts' song of the luckless roughneck and the anonymous "On the Allegheny Shore." Roberts sang his song to the melody of "Frankie and Johnny."[6]

> Come listen all you rounders;
>   I'll tell you a tale that's sad.
> All about a poor roughneck,
>   And the troubles that roughneck had.
>   He's a good old boy,
>   But he's broke again.
>
> Found him a mudder,*
>   And he asked for a meal.
> Mudder said, "Where have you been working at?"
>   And he told him in the oilfields,
>   " 'Cause it's all I do,
>   'Cause it's all I do."
>
> The mudder bought him ham and eggs,
>   Said, "You bet I'll feed you well.
> You're the very man I need right now;
>   I'm putting down a wildcat well.
>   I'm one man shy,
>   And I got to get down."
>
> So he screws that pipe together
>   And he runs down in the well.
> Says, "If I ever get on bottom again,
>   I'll rest a little as sure as Hell.

*"Mud engineer," i.e., a rotary driller.

I'm about all in,
I'm about all in."

When he hits on bottom,
    The driller turned it round and round.
"Down with the hole, boys.
    Got to take another core, boys.
Gotta see what we've found.
        I believe it's sand.
        I believe it's sand."

"We made three round trips last shift,
    And I did three more today."
Driller said, "We gotta get an oil show,
    Or we'll never gonna get no pay.
        He's a hard old guy;
        He's got no heart."

"Oh, boots all worn, and leakin'
    And I ain't got on no socks."
Said, "I wish that boy would learn to roll a bit.
    Put a runner on these hard rocks.
        It's a roughneck's friend.
        It's a roughneck's friend."

For he's all wet and muddy.
    Chain tongs they won't hold.
He tries a straight job as a fireman,
    But the fireman says it's too darn cold.
He's to stay out there where it's good and warm.
    He gets his sleep at night time,
In a little old wooden shack.
    Bed's all damp, and the blanket's thin.
And the wind comes through those cracks.
        Oh, Lord it's cold.
        Oh, Lord it's cold.

"Light bread is gettin' moldy,
    And the ham done gettin' stale.
We'll have prunes again for breakfast, boys,
    For it's never known to fail.

At the boarding house in the oilfields,
  Oh, the eggs come all from cold storage,
And the bacon from a poor old sow.
  Hot cakes make good gaskets, boys,
And the butter never saw no cow.
        It's the same old thing,
        At the boarding house."

He finally makes a payday,
  And he beats it into town.
Got no particular business there,
  Just going in to look around.
        See what he can see,
        See what he can see.

Bar tender gives him a couple of drinks.
  Pretty soon he buys a quart.
He starts in spendin' his money then,
  'Cause he don't want to be called short.
        He's a good old boy,
        But he don't know much.

Woke up before day next morning,
  Sleeping on a old hard bed;
Run his hands down his pocket, boys,
  But he hasn't got a red.
        He's broke again,
        He's broke again.

Says, "So I walk over to the window,
  Just to look out and see the stars;
All I find at the window, boys,
  Is a 'row of them iron bars."
        He's in jail again,
        He's in jail again.

"So, I guess I'll be seeing the judge this morning,
  As usual along about ten.
He'll look down over his glasses then,
  Say, 'Boy, you've been drunk again.
        It's the rock pile now,
        It's the rock pile now.'

"But I'll take my old sledge hammer,
   And I'll give those big rocks Hell.
I'll rather be working for the county, boys,
   Than drilling on a wildcat well.
      The old jail's home,
      And the hash ain't bad."

"On the Allegheny Shore," published anonymously in the
Oil and Gas Man's Magazine for January, 1911, has the characteristics of a folk ballad, but I have found no evidence that it
either came from or entered the oral tradition. Hermes Nye
included it, under the title of "The Dying Toolie," in his *Texas
Folksongs* (Folkways Records f.P 47-1), singing it to the melody
of "The Dying Hobo," which is clearly suggested by the opening
lines.

On The Allegheny Shore; or The Dying Toolie

O, the toolie of a drilling crew
   With short and feverish breath
In a boarding house in Beaumont
   Lay near the point of death.

His driller stood beside him
   While his life did ebb away
And bent with pitying glances
   To hear what he did say.

The toolie's lips did tremble
   As he took the driller's hand
And said, "No more will I temper bits
   To pierce the Texas sand;

"No more the boiler I will fire
   Nor climb the derrick high;
No more I'll eat a midnight lunch
   Beneath the southern sky.

"A message I would have you take
   To my pleasant home of yore

For I was born in Franklin
   On the Allegheny shore.

"Tell my father not to shed a tear
   Nor bow his aged head
When you gently break to him the news
   That his eldest son is dead;

"That the precepts of his teachings
   I have ever kept in mind
And never skipped a board bill
   Though it took my hard-earned dimes.

"Tell my mother that her loving son
   When death was drawing nigh,
Looked forward to a brighter home
   And did not fear to die;

"That in my dreams I saw her form
   Beside the cottage door
In that far-off home in Franklin
   On the Allegheny shore.

"There's another, not a sister,
   Who lives in Olean;
I've not seen her since that bright June day
   She pledged to me her hand.

"Well I can see the rolling river
   Where together we have strolled
Along its shady, mossy banks
   In those happy days of old,

"As we talked of the bright future
   That time held for us in store,
Of a home in dear old Franklin
   On the Allegheny shore."

The toolie's lips were silent
   There was one angel more;
Saint Peter wrote "Tim Murphey
   From the Allegheny shore."

The one ballad that did have wide oral circulation, particularly in the older oil states, was apparently never sung, but many old-timers have heard it declaimed in boardinghouses and saloons. I have had access to two texts, one published by Margaret Flanagan in the *New York Folklore Quarterly* for May, 1945 (I, 88-98), the other supplied by Harry M. Rhinelander of St. Louis, son of a Pennsylvania drilling contractor. He thought it was written by one Ella Hanrahan, a boardinghouse keeper. Miss Flanagan's informant, Omar McQueen of Scio, New York, had a handwritten copy which his wife had made from a printed pamphlet, he thought in the late nineties. His copy did not show the name of the author, but he thought it was Ella Handerhand. A note in the *Oil and Gas Man's Magazine* for August, 1912, however, attributes the ballad to Mrs. Charles A. Muller, the wife of a pipeline employee.

The textual differences between the two versions are not sufficient to require the reprinting of the seventy-seven stanzas here. There are some changes in place names, reflecting no doubt the experience of the declaimer, and changes in phrase of the sort that commonly result from oral transmission; for examples, "These I did recall"; "In grief I did recall"; "I thought was very dim"; "I thought was slim."

A driller of Irish descent has a dream in which

> An Angel clad in robes of white,
>   "Come go with me," did say,
> "For yours shall be eternal night,
>   Or never ending day."

In "Heaven's air" he sees a flag upon which are written the words, "No Irish need apply." To Saint Peter he confesses that he is a driller and that his name is Dennis Ragen, but says that his mother (who was really a Murphy) was a Jacobs of the Pennsylvania Dutch. But Saint Peter, detecting the lie as well as the reason for it, points out that the sign applies to hell. Ragen recounts his life as a driller, confessing that away from home he passed as a single man, because at the boardinghouses

> The married men got bread and beans,
>   The single men got pie.

He tells of the hard life he has lived, how

> I have breasted Bradford's frozen snow
>   On many a wildcat well
> And climbed the derrick oft at night
>   When rain in torrents fell.
> Through lightning, tempest, flood, and storm
>   I've bravely fought my way.

Saint Peter listens, saying, "Blessed are they that labor/ And bow before God's will."

> "A driller's life I know is filled
>   With crosses and with care
> But beyond the grave he's richer
>   Than any millionaire."

At this point trains and an ocean steamer begin to arrive filled with drillers from all the oil regions.

> Then up the flowery path we marched
>   With band and banners gay,
> We rang the bell at heaven's gate;
>   What did Saint Peter say?

> Don't ask, for just then
>   The old clock struck eleven
> And I awoke to find that I
>   Was far away from heaven.

If it had been possible to find some of the Negroes who worked in the Texas coastal area at the beginning of the century, this chapter would no doubt have been longer. Until animals were supplanted by trucks and motor-powered earth-moving machines, the teamsters who hauled supplies, dug earthen storage pits, and threw up embankments around tanks were chiefly Negroes. Plummer Barfield remembers that they sang their commands to their mules, and along about quitting time, especially, they would sing songs.

When the Negroes would go to singing, why, the mules would go to braying because it was pretty near sundown.... And they made their songs as they went, and pretty muchly some of them was the same tune ... that they sing their spirituals [to] right now. And some of them were singing spiritual songs, no doubt. And some of them were using other language that could have been omitted.[7]

Barfield does not remember any of the songs, and it is doubtful whether they can at this date be recovered.

14

Tall Tales

THE BEST-KNOWN tall tales of the oil industry center around three heroes: Paul Bunyan, Kemp Morgan, and Gib Morgan.

PAUL BUNYAN

When and where and how Paul Bunyan made his first appearance in the oil fields is an unsolved problem, and one that at this date is probably beyond solution. He first appears in print as an oil field hero in two articles of independent origin published in *Follow de Drinkin' Gou'd,* the Texas Folklore Society Publication for 1928, "Paul Bunyan: Oil Man," by John Lee Brooks, and "Pipeline Days and Paul Bunyan," by Acel Garland.[1] Brooks's introduction to Paul Bunyan came in the summer of 1920 in the Hewitt field near Ardmore, Oklahoma, where as a boll weevil, or green hand, he heard many casual references to the "miraculous time and labor-saving devices that the powerful and ingenious Paul Bunyan used on his rig," but he found no cycle of stories, no extended tales. In the summer of 1921 Brooks worked as a tool dresser in the Breckenridge, Texas, area. His driller and the one on the opposite tour (shift), both West Virginians, used to kid him by calling to mind some of Bunyan's exploits. For example, they would tell him what Bunyan did to a boll weevil who spliced a rope exactly as Brooks had spliced one.

Two years later Brooks went to the fields as a collector of folklore. He worked in the area around Ranger, Cisco, Breckenridge, and Eastland, and at Big Lake and McCamey. "I found," he reported, "no one who gave me more than a reminiscent smile

and perhaps a slight incident or two . . . my efforts resulted in a
heterogeneous mass of incident that spoke more often of Paul
Bunyan as a rig builder and driller, but also as a pipe liner, a tank
builder, and even as a constructor of telegraph lines."[2]

As a rig builder Bunyan "could sight so accurately that no
plumb line was necessary." He could build a rig in one day by
himself. He used an eight-pound hatchet and drove the large
spikes to the head with a single blow. He could build a pair of
bull wheels in a half day and could hang the walking beam
single-handed. If anything fell from the crown block at the top
of the derrick, he caught it, thus saving both material and the
heads of the workers. Once he built a rig and spudded in with a
Ford motor. He boasted that he could dig faster with a sharp-
shooter (i.e., a narrow spade) than a crew could with a regular
rig. Nobody called his bet, so he was not put to the test, and thus
had no opportunity to become a John Henry. Sometimes he would
drive his sixteen-pound sledgehammer in the ground and get oil.
Once when he was on the derrick he threw his hatchet at a man
below, who had angered him. He missed the man but the hatchet
penetrated the oil sand. He cased the hole and made a million
dollars from the oil. He invested this money in Mail Pouch tobacco,
which he intended to soak in whiskey and sell. "His own appetite
got the better of him, however, and he chewed it all himself."

He built a rig so big that he had to have telephones installed
to communicate with the derrick man, who came down only twice
a month on payday. He built another that reached to heaven,
where the crew lived until the well was completed. Once he drilled
to China. Another time he was blown from a derrick at Bakersfield,
California, five hundred miles into the Pacific Ocean. He was
attacked by a whale, but fortunately he still had his hatchet in
his hand. He killed the whale, mounted it, and paddled home.

While working with a rotary rig, he invented flexible drill pipe
which could be wound around a drum, thus saving the many
hours of labor required to raise and lower the drill pipe in sections

when a bit had to be changed. This pipe, however, was for others. On his own wells, he merely ran the pipe up into the air and held it while his helper changed the bit. His boilers were so large that workmen had to be careful to avoid being sucked in by the injectors. Once, on a cable rig, his tools got stuck in the well. In attempting to jar them loose, he raised up the whole lease. In Brazil he struck a vein of rubber which flowed out and covered the derrick floor. His tool dresser fell from the derrick, struck the rubber, and began bouncing. Paul finally had to shoot him to keep him from starving to death.

Once when he was lowering the tools in the well, he got caught in the steel cable and was carried to hell, where the devil showed him the sights, including his harem. Paul took a fancy to one of the beauties and tried to take her. The devil chased him back up the well. Sometime before, a roustabout had cut off one of Paul's legs and thrown it in the well. He saw it in hell, but the devil would not let him have it.

He shot a well with five hundred quarts of nitroglycerin. It immediately started to flow and Paul sat on the casing head to save the oil. The casing shot up into the air, and Paul had to sit there three days while a derrick was being built to rescue him. He pulled up dry holes and sold them for postholes.

As a pipeliner he built a buttermilk line to supply his crew. His tongs were so heavy that it took four men to carry them.

Garland reports some of the same incidents. He tells also of the cattle pipeline Paul built to Chicago. It brought about great savings in the shipping of cattle, though some of the calves and yearlings got lost in the threads. In casing a well, Paul would lay out the casing on the ground. Then he would end it up and shove it in the well. Once when the bit was not cutting the formation, Paul grabbed the drill pipe, raised it, and threw it down so hard that it broke the rock and pulled the derrick in after it. Another time he drilled successively into formations of cornbread, buttermilk, and good old turnip greens. He invented a perpetual motion drilling rig. The tools were made of rubber, and once they had been set in motion, they bounced up and down without further attention.

These incidents are representative of the legend of Paul Bunyan in the oil fields. My own field work and the tape recordings made under the auspices of the University of Texas in 1952 and later reveal a few variant incidents, but make no significant addition. The material is even more fragmentary than Brooks found it in the 1920's. Nor was there any interest in Bunyan among either the retired men or those still active. One man who began work in the oil fields in 1904 said there used to be too many Bunyan stories. They went in one ear and out the other. "One joke after another. Like Paul Bunyan out there fishing; a big fish story. He . . . caught a whale, and when he got him up to the bank, why the whale got Paul. So that was the last of Paul Bunyan at our place."[3]

Another man, who began work in 1901, could remember that he had heard so many Bunyan stories that he "never paid any attention to them."[4] He could not recall any. A third man, whose career in the oil business began in 1916, could recall only that it took thirty-three flatcars to haul Paul's Kelly joint.[5] These responses are typical. They indicate that Paul Bunyan does not belong to the living folklore of the oil industry.

Brooks thought Bunyan had been brought to the Southwest by oil workers from West Virginia, many of whom came to Texas and Oklahoma. But in the light of further investigation, this assumption seems untenable. Of the many retired oil workers I have interviewed in and from the older oil-producing states—Pennsylvania, West Virginia, Ohio, and Kansas—not one has heard of Paul Bunyan in those states.

It is not impossible that Paul Bunyan's migration from lumber to oil took place at Beaumont soon after the discovery of oil there in 1901. Aside from rice farming and cattle-raising, lumbering was the chief industry of the region. But there is no evidence that Bunyan was known to the lumberjacks and sawmill workers. They

were chiefly Negroes, and there was little contact between them
and their counterparts in the North.

Nor is it possible to date with accuracy Bunyan's arrival in the
oil fields. No informant whose service dates to the beginning of
the century can say with certainty when he first heard of Paul
Bunyan. But the memories of two oilmen place the date early
enough to indicate that he came by way of oral tradition. A. P.
Hodge believes that he had heard of Paul Bunyan before he moved
from South Texas to the Wichita Falls area in 1913, and that
stories of Bunyan were current in the coastal region as early as
1910.[6] His memory is supported by that of H. A. Rathke, who
heard of Bunyan in South Texas "somewhere close" to the Spindletop boom. But he would not attempt to fix a date. "I don't know
when he got to Texas," Rathke said. "He was the biggest driller
that got to Texas, though, when he did get there."[7]

KEMP MORGAN

A number of writers have made the hero of the tall tale of
the oil fields not Paul Bunyan, but Kemp Morgan. In his introduction to *Here's Audacity,* published in 1930,[8] Frank Shay defines
all American tall tale heroes as "audacious industrialists." They
are, he thinks, one man whose name and face vary from locality
to locality and from occupation to occupation. He is Paul Bunyan
in the Northwest, Tony Beaver in West Virginia, Pecos Bill in
the Southwest, John Henry in Virginia, Casey Jones on the railroads, and Old Stormalong on the old windjammers. "In the oil
fields of Texas and Oklahoma he is a rotary well-digger known
as Kemp Morgan," to whom a chapter is devoted.

Morgan's adventures as told by Shay are based directly upon
the articles of John Lee Brooks and Acel Garland, presented with
hopeless confusion in both geography and technical processes. On
the "plains" of East Texas, where Kemp once camped for the
night, the sand dunes shifted and left his pack mules, upon which
he carried his drilling equipment, hanging forty feet in the air.

Whether it was a similar sandstorm that moved Smackover out of Arkansas is not clear. At any rate Bull Cook Morrison— borrowed from Garland—who told Shay about Kemp Morgan, worked in a restaurant in Smackover, Oklahoma. Kemp is called a rotary driller, yet when he made preparations to drill, he mounted bull wheels and hung a walking beam, articles not found on rotary rigs. He invented a drum around which "steel drills" rather than drill pipe could be wound. Although he mounted bull wheels and hung walking beams *before* drilling, for some reason he built his derrick *after* the wells had been brought in. When he located oil, "he would take a sharp-shooter, a long handled shovel [they are not the same tool] and begin spudding in."

He "could sight so accurately that no plumb line was necessary." He could build a full-sized derrick in a few hours. He drilled into an alum mine and into a formation of rubber. Once he brought in a gusher when the weather was so cold that the oil froze as it came out of the well. He sawed it into sections and shipped it on flatcars. He sawed up dry wells and sold them for postholes. When a boiler blew up, he mounted it and rode it back to earth. He drilled a well fifty thousand feet deep and when it started to flow, all the dusters in the country around began gushing too. The derrick was so high that he had to hire thirty men in order to have one on duty in the derrick. At all times there would be fourteen men on the way up, fourteen on the way down, and one on the ground. After Kemp opened the Smackover field, he sold his wells for a million dollars and invested his money in Mail Pouch tobacco. This was a bad investment, for "his own lousy appetite got the best o' him an' he chewed it all hisself."

The one talent that Kemp had that Paul didn't have was his ability to smell oil, no matter how deep. It was with his nose that he located the oil sand fifty thousand feet deep.

Carl Carmer in his *The Hurricane's Children*, 1937,[9] includes a redaction of Shay's chapter on Kemp Morgan, and Olive Beaupre Miller elaborates it in *Heroes, Outlaws, and Funny Fellows of*

American Popular Tales, 1942.[10] Carmer's Kemp Morgan, like Shay's, drills his wells and then builds his derricks. Olive Miller introduces Bill Cook Morrison as the informant, tells about the sandstorm, and has Kemp drill a well ten miles deep, which is near enough fifty thousand feet, and sell his properties. She does not say what he did with the money. People from all over Oklahoma sent for him to smell out oil, and as a result,

Men, women, and children who had been living in shacks found themselves, thanks to him, able to buy the things they had longed for all their lives.... Children got toy trains, dolls, or bats and balls; women got fine fur coats, electric washing machines, or beautiful dresses; while the men bought radios and shiny big automobiles.

The children said Kemp Morgan was better than Cinderella's godmother.

Clearly Olive Miller's Kemp Morgan is a symbol of the beneficent wildcatter, who goes about increasing the wealth of the land and thus enabling the American people to buy the gadgets that make them happy.

In 1940 J. H. Plenn introduced both Paul Bunyan and Kemp Morgan.[11] He speaks of round robin stories oil workers tell each other in their hotel rooms on their off days in Houston, with "Kemp Morgan, the folk hero of the oil fields, and Paul Bunyan, the lumberjack giant, sharing honors.... From the various bits, including my own extemporaneous contribution, I pieced together the story of Paul Bunyan digging oil wells in Texas."[12] Kemp, visiting Paul in the north woods, finds him in distress because the grease in his watch has caked up and it has stopped running. Kemp suggests that he come to Texas, where there is plenty of "rock oil" to keep the watch lubricated. They form a partnership and begin operations. Paul strikes oil by driving his ax into the ground; he tries to cap a well by sitting on the casinghead and is blown high in the air; he invents pipe stretchers; he and Kemp drill a well to China, which produces chow mein and chop suey, after passing through such strata as sowbelly, watermelons, Mexican blankets, and rubber, from which they fashion such articles as tires, boilers, and printer's type.

In 1959 Mrs. Ella Lane Carl published a book entitled *The Letters of a Texas Oil Driller's Wife*.[13] In a letter dated from Pampa, Texas, January 25, 1931, she writes, "Here are some tall tales about the oil drillers and their large scale operation." It is evident that the epistolary form is a literary device, and that this letter was written long after the date indicated, for the tales she introduces are taken almost verbatim from Plenn. The quotation of a paragraph from each will suffice:

Plenn: The deepest well they dug was Old Fifty-Seven. It went right on down to China. It was a gusher, too, but no oil came out of it. After the tools had blown out, a lot of white stuff started coming out. It was rice. Then there was a rush of wind, the rice stopped flowing. Soon there came a stream of hard, crisp fried noodles, and these were followed by a long gurgle and escape of steam, and then a steady, pulsing flow of chow mein and chop suey. This came out over the noodles and stacked up into big mountains.[14]

Carl: The deepest well Paul and Morgan ever dug was old "Fifty-Seven." It went right down to China and a gusher it was too. But no oil came out of it—only rice. Then there was a rush of wind, the rice stopped flowing and soon there came a stream of hard crisp, fried noodles and these were followed by a long gurgle and escaped steam, and then a steady pulsating stream of chow mein and chop suey.[15]

Obviously Mrs. Carl furnishes no evidence that the oil workers at Pampa were telling Kemp Morgan stories in 1931.

Bob Duncan's story of Kemp Morgan in *The Dickey Bird Was Singing*,[16] 1952, is largely his own invention. He follows Shay in giving Kemp "the greatest nose on earth," and Miller in stressing his benefactions. These consist in part in his paying huge dividends from his own fortune to the victims of the swindling promoter of the Pearson Oil and Investment Company.

In introducing his hero, Duncan says, "More has been written and said about him, perhaps than any other man in the history of oil." In reality what has been written about him, for all its repetition, would not fill a book, and what has been said about him is less. I have yet to encounter Kemp Morgan in the oral tradition. Not one roughneck, driller, producer, not one person connected with the production of oil have I found who knew of him as an oil field legend.

In writing on Bunyan Shay had commented upon the tendency of occupational heroes to invade each other's provinces. He denied that Bunyan went to the oil fields. "That's another chap named Kemp Morgan." I suspect the name was Shay's invention, and its similarity to that of another hero was coincidental.

When I ask people from the oil fields about Kemp Morgan, a frequent reply is, "Don't you mean *Gib* Morgan?"

GIB MORGAN

Gib Morgan, unlike Paul Bunyan and Kemp Morgan, was a man of flesh and blood; he had been born on a certain day and at a certain place; he had been a soldier and had fought in certain battles in a certain war; he had married; he had had children; he worked in the oil fields of many states; he retired and lived in homes for disabled veterans; he died, leaving an imprint on the traditional lore of the oil industry.[17]

He had grown up with that industry. He was born in Clarion County, Pennsylvania, in 1842, seventeen years before Drake's discovery. When the Drake well came in, he was living at Emlenton, less than forty miles from Titusville. Before the discovery of oil his surroundings were essentially those of a frontier culture. Population was sparse, but even so there were more people than a primitive economy could well support. In their poverty and isolation they possessed a rich and voluminous folklore, which Gib Morgan early assimilated. They sang old ballads; they passed on to the children an oral tradition about Washington and Half King at Venango, about the Whisky Rebellion and Indian massacres and captivities. There were tall tales about hunting and fishing. Gib's father, George Morgan, was said to have been a good storyteller.

George Morgan was considered well-to-do by his neighbors. His barge works at Emlenton, later moved to Tionesta, prospered and he died a respected member of the town council.

Thirteen days after Fort Sumter was fired upon, Gib, not yet twenty years old, enlisted in the Tenth Pennsylvania Volunteers. During weeks of uncertainty and confusion and epidemics of sickness, Morgan became known as "the best storyteller in the regiment." This was about all the distinction he was to gain. Although the Tenth was to see plenty of action and to be twice praised in the dispatches of General Meade, Gib Morgan never advanced beyond the rank of private.

When he got home from the war in June, 1864, oil excitement was reaching the crest of the first speculative boom. Visitors to the region in 1864-65 write of "sharp-eyed speculators from New York, Philadelphia, and Pittsburgh, carpet sack in hand . . . going down to secure 'a big thing'; traders anxious to open a line of custom; rough fellows going down to work at the wells; and old farmers . . . who had been made rich by farms that had previously made them poor."

Gib went to work and in due time became a driller. In 1868 he married Mary Elizabeth Richey of Richland on Richey's Run. The oil field was no place to take a bride, so Gib left Mary at Tionesta, where his father had just moved. Mary Morgan died in 1872 after giving birth to her third son, Warren. Gib used to say that his luck changed when Mollie died. He was never able to settle down. He became a boomer, or as one of his friends put it, a "gypsy driller," and spent the next twenty years roaming from field to field.

Wherever he went he was known for his wit as well as his

tales. As he was passing one day in Macksburg (Ohio) on his way to tour, someone said, "Gib, tell us a lie."

With a show of deep emotion, Gib replied, "I can't tell you a lie now. I've just got word that the cable clamps have slipped and killed my brother at the well."

Deeply humbled, the men expressed their regrets and Gib went on. After him came a crowd to find out about the accident. They found Gib calmly at work.

"You told us your brother had been killed," they accused.

"You asked for a lie, didn't you?" Gib replied.

Gib was present but not working when the famous Greenlee and Frost well came in at 25,000 barrels at McDonald in 1891. The crew and extra help were trying to close the well in. "Mr. Greenlee," writes an eyewitness, "came up through the woods toward the well with an umbrella over him to keep the spray off of him. Gib was sitting on a stump watching the workers. He had on a hickory shirt and jean pants. Mr. Greenlee says, 'Gib, why don't you go to work and help close in this well? We're paying a thousand dollars an hour.' "

"Gib raised up and said, 'Mr. Greenlee, do you suppose I could spoil these clothes for a thousand dollars an hour?' "

Gib was in Warren County during the Cherry Grove excitement of 1882. He was hired by a pipeline foreman to clear the ground and grade for a tank farm. The foreman was a driver of labor who liked to impress his men with his toughness. Once when he boasted that he had killed a man, Gib told his famous tale about his fight with a Negro. Whether this was the first time Gib had told the yarn, I do not know. Many of my informants have thought that they were present when Gib "first made up" a certain yarn.

Gib retired from actual drilling in 1892. That year he applied for and secured a Civil War pension. He lived with his son Warren at Eau Claire for a year or two, then in 1894 was admitted to the National Home for Disabled Soldiers at Marion, Indiana. In 1901 he was transferred to Danville, Illinois, and in 1904 to Mountain Home, Tennessee. When D. M. Hosack was living near Marion in 1901, he used to visit Gib. What he remembers vividly is a thin, wiry man seated on the lawn surrounded by some two hundred other veterans listening to his yarns. Every summer Gib would get a furlough and go back to the oil country of Pennsylvania. His annual homecoming became an institution, almost a triumphal entry. His first stop would be Pittsburgh. There he would register at the Bowyer Hotel, long the favorite hostelry of the oil fraternity. The *Pittsburgh Gazette,* which had featured oil news from the beginning, would soon learn of his presence (no doubt with Gib's co-operation) and would publish a story in which Gib would be linked with the first robin of spring. Old friends would drop in, along with others who knew him only by reputation, eager to hear his tales. But Gib would not be rushed. Somebody would propose a drink. They would file out to the bar. Bottles and glasses would be set out. After a few minutes of decorous reticence, Gib would launch out into his adventures in Pennsylvania, West Virginia, Ohio, Indiana—places he had been—and in Texas, India, Baku, the Fiji Islands, places he had never seen.

After a few days in Pittsburgh he would go on to Emlenton, Franklin, Oil City, Titusville, to see old friends, and on to Eau Claire where his son Warren lived. His pension was never more than twelve dollars a month, and sometimes he had difficulty in making his money hold out until his furlough was over. But even when it was all spent, he could manage very well. It was understood that when he drank with the oil people, the drinks were on somebody else. If any of the newcomers were ignorant of this custom, Gib had a delicate way of informing them. He would tell them about the *whickles.*

The whickles were strange petroleum-eating insects that were responsible for the declining production in the Pennsylvania fields. Gib was doing his best to exterminate them. His method was to sprinkle applejack on the bushes around the wells; the whickles would come and get drunk and Gib would catch them. But apple-jack cost money and that was the reason Gib was always broke.

One winter Gib obtained board and lodging in the same way he did his drinks. In the fall of 1897 John C. Chambers opened up a general store in the new oil field at Klondike. Gib got to coming to the store, and customers followed him. Chambers paid his room and board from November to June to have him around the store as an added attraction.

In the spring of 1908, Gib did not show up in Pittsburgh, and in the following February word came of his death at Mountain Home. By that time, his tales were being repeated by hundreds of oil field people.

Not all his tales are concerned with the oil industry. He would, for instance, tell a Munchausen tale with a Russian setting, saying it happened to him when he was drilling in Baku, or he would tell tales of hunting, fishing, and planting that he had no doubt heard as a child. He told of a remarkable horse he had once owned twenty-two yards long with three speeds forward and reverse, which served him well both on the road and on the racetrack. But his best tales were those of the oil fields.

They are distinguished from those of Paul Bunyan and Kemp Morgan by their greater elaboration and by their absence of super-naturalism. The more fully developed versions come from men who knew Gib personally and had heard the tall tales more than once from his own lips. Gib was not a giant and he made no pre-tense to the mastery of cosmic forces. This may have been a result of the limitations he placed upon his own imagination in making himself the hero of his tales. Since the hero was actually present as narrator—present with his hundred and fifty pounds of weight and his five feet, nine inches of height—he could not be endowed with the size and strength attributed to David Crockett and Paul Bunyan. He did invent a tool dresser twenty-eight inches between the eyes and tall enough to grease the crown pulleys at the top of the derrick without taking a foot from the ground. But he made little use of the giant except to kill him off. Similar limitations applied to Gib's business ventures. Since he was obviously not a wealthy man, his tales had to explain his failures. This sort of explanation, incidentally, is one of the most common themes in both the history and the folklore of the oil industry.

The tales below are representative but not exhaustive of the Gib Morgan canon.

Gib Morgan's Fight

Gib Morgan was a peaceful and law-abiding man whose motto was "Live and Let Live." He tried to get along with everybody, and he generally did. In fact during his long career as a tool dresser, driller, pipeliner, production superintendent, and what not, he had only one fight and he wouldn't have had it if he could have helped it.

Gib was working for a pipeline company bossing a gang of nigger ditch-diggers along the Ohio River. They were all fine hands but one. This one was a big buck who thought he knew more than Gib. He wouldn't do his work right, and when Gib would jack him up, he would give him a lot of back talk. Gib saw that he would have to curb him a little bit or he would ruin the whole gang. So one day when he was particularly sassy, Gib corrected him with a pair of forty-eight-inch pipe tongs.

Well, that nigger grabbed Gib around the waist and they both rolled into the Ohio River and sank to the bottom. The next thing Gib knew the nigger had drawn a knife and started to whittle on him. Gib had to protect himself. He drew out his jackknife and started to work on the nigger. Then began the most God-awful battle that ever took place in the United States.

All the men on the gang stopped work and lined up along the river bank to watch the fight. The news spread and more people came. The railroads ran excursion trains to the scene, and all the steamboats stopped and tied up. The farmers on both sides of the river charged for standing room and cleared over fifty thousand dollars. Tens of thousands of dollars were bet on the outcome of the fight.

The only way the spectators had of knowing how the battle was going was by watching for the pieces of flesh that came to the surface of the water. When there were more pieces of white-skinned flesh than black-skinned flesh coming up, the odds were in favor of the nigger. When there were more pieces of black-skinned flesh than white-skinned flesh coming up, the odds were in favor of Gib. Not all the pieces, however, came up. When the catfish found out what was going on, they gathered from miles up and down the river for a big feast. Every once in a while one would swim between Gib and the nigger and get cut to pieces.

They fought and they fought until finally their knives got so dull they wouldn't cut any more. They agreed to a truce so they could come up and grind their blades. When they came out they discovered they had been fighting for two weeks, and they were powerfully hungry. They went to a restaurant and each ordered a beefsteak four inches thick. When they had finished eating, they felt so good they called the fight off.

How Gib Saved a Farmer's Life

When the oil excitement broke out in West Virginia the Scarcely Able and Hardly Ever Get Oil Company sent Gib in to drill a well. Gib thought he had seen hills in Clarion and Verango counties back in Pennsylvania, but by the time he got to Wheeling, he realized that he had not known what a hill was. At the place where he boarded, he could look up the chimney and see the cows come home. He drove the stake for the first location in a corn patch because that was the only place he could find a piece of level ground big enough to set up the rig on. From the top of the derrick he could look down the chimney and see the farmer's wife churning.

Gib had some bad luck on that first well, which turned out to be the good luck of the farmer. About five hundred feet down he broke a pin and lost his bit. The only thing to do was to rig up a string of fishing tools and go to work. He put on a pair of long stroke jars and a horn socket. The horn socket was a tube with a conical flare at the bottom to fit over the broken bit and a spring latch to hold the bit as it was being drawn from the well.

As Gib was making up the string of tools he noticed the farmer out hoeing corn. At the end of each row he would stop and lean on his hoe and rest a while before he hoed the next row. Just as Gib was about to lower the tools into the well, he heard a loud crack like a stick breaking and saw a cloud of dust coming from the edge of the field. At first he thought it was a hog going down the hill, but pretty soon he heard yelling and cussing, and a boy hollered out, "Ma, Ma, Pa's done fell out of the corn patch."

Gib knew that something had to be done and done quick. He threw the tools off the ledge and let out cable. When he had run out about fifteen hundred feet, he caught the old man's head with the horn socket and reeled him back up.

That night the old lady said, "Jim, I told you you ought to build a fence around that there corn patch. It's dangerous to work up there."

The old man said, "It warn't a fence I needed. It war a new hoe handle. I knowed that old one war a-gittin' weak."

The old man and the old lady surely were grateful to Gib for saving his life. After that nothing was too good for him.

How Gib Drilled on Pike's Peak

The only other place where Gib ever drilled in country as perpendicular as West Virginia was in Colorado. He was then working for Standard Oil. They had located a well just by making a cross on the map, and Gib was sent in to drill it. When he got out there, he found that they had put that cross right smack dab on top of Pike's Peak. The crew wanted to set up down in a valley somewhere. They said the brass hats back East wouldn't know the difference anyway. But Gib said no. He hadn't seen a location he couldn't drill on yet, and he wasn't going to be stumped at his age.

So they snaked the timbers up and built a derrick on top of Pike's Peak. But when they got the rig up, there wasn't any room for the engine and boiler. The nearest piece of level ground big enough to put them on was twenty-three miles below. It took forty-six miles of belting to connect the power plant, and a belt that long will stretch a good deal. They had to relace it every few days to take up the slack. Gib saved all the pieces they cut off and had enough leather to keep his boots half-soled for the rest of his life.

It was too far to walk from the engine to the rig, so Gib bought a mountain mule to ride back and forth. At first he was a little bit leery about riding the mule down, but the natives said there wasn't any danger. All the mules in that country were used to mountains. They were surefooted and never stumbled. Thus assured, Gib got on the critter to ride him down to the engine. As he rode he could see the mule's head between the stirrups.

When he got about halfway down he felt something warm on the back of his neck. He ran his hand under his overcoat collar and when he drew it back it didn't smell like Hoyt's Cologne. He decided that while the mule might be safe enough, he would prefer to get about some other way. He sold the brute, and after that when he wanted to go from the engine to the rig or from the rig to the engine, he just threw his saddle on the belt and rode it up or down.

Gib's Biggest Rig

Once Standard Oil sent Gib down to Texas. There was a certain place down there where they figured that there was oil under the ground, but they hadn't been able to get to it. They had sent their crack drilling crews and production men down, but the formation above the oil sand was so cavy that they hadn't been able to make a hole. They would start with a twenty-four-inch bit and case with a twenty-two-inch casing. Then they would make a few more feet of hole and would have to set a twenty-inch casing. They would cut a little more ditch and then they would have to case again. And it would go on like that until the casing became too small for the tools to go through, and after all that expense they would have to abandon the hole.

Finally John D. himself called Gib in and showed him the logs of all the wells they had tried to make, and said, "Gib, do you think you can make a hole down to the oil sand?"

Gib looked at the logs a while and then he said, "John D., if you'll put up the money, I'll put down the hole."

"It's a deal," said John D., and they shook hands on it, but they didn't drink, both being temperance men.

First Gib went over to Pittsburgh to see the Oil Well Supply Company and told them how to make the special tools he wanted, some big tools and some little tools. Then he went to Texas and started putting up the rig.

The derrick covered an acre of ground, and since Gib expected to be there for some time he fixed it up nice. He weatherboarded it on the outside and plastered it on the inside. It was so high that he had it hinged in two places so that he could fold it back to let the moon get by. It took a tool dresser fourteen days to climb to the top to grease the crown pulleys. That is the reason Gib had to hire thirty tool dressers. At any time there would be fourteen going up, fourteen coming down, one on the top and one on the ground. A day's climbing apart he built bunkhouses for the men to sleep in. These bunkhouses had hot and cold showers and all the modern conveniences.

By the time the derrick was up, the tools began to arrive from Pittsburgh. The biggest string of tools reached to within ten feet of the crown block. The drill stem was twelve feet in diameter. At the first indication of caving Gib cased the well with thousand-barrel oil tanks riveted together. This reduced the hole to twenty feet. He put on an eighteen-foot bit and made about fifty feet of hole before he had to case again. Down about five hundred feet he had to go to a smaller bit, one about six feet in diameter. At a thousand feet he was using standard tools. At two thousand feet he was using his specially made small tools and casing with one-inch tubing. But he hadn't figured it quite fine enough, for he hadn't got to the oil sand when the smallest drill he had wouldn't go through the casing. But that didn't stump Gib. He brought in the well with a needle and thread.

How Gib Lost a Fortune

Some of Gib's most remarkable adventures took place in the Fiji Islands, where he was sent by a British syndicate to drill for essence of peppermint, which was expected at a depth of nine thousand feet. If he failed to get a paying well at that depth, he was to drill on into a lower sand for bay rum, which was expected at a depth of twelve thousand feet.

He set up his rig on one of the smaller islands, spudded in, and soon had the drilling under way. The going was easy and he had a good deal of leisure time on his hands. He knocked around on the island a good deal looking for companionship, but he didn't find the natives very sociable, and they couldn't speak United States anyway. Finally he got so bored that he felt he had to get away for a day or two, and as the work was coming along fine, he visited the port on one side of the larger islands. There he met the Noble, Grand, Worshipful Master, Chancellor, Commander, Exalted Ruler and Jibbonancy of the Islands. This potentate, who spoke Oxford, gave Gib the keys to his spacious grass and palm-leaf mansion and entertained him so royally that he forgot all about the well for a whole week.

When he got back to the location, drilling had stopped. The men were sitting around under palm trees, slick and fat and too lazy to move. They had struck a flow of buttermilk soon after Gib left, and they had done nothing the whole time he was gone but sit around and drink buttermilk and sleep.

Gib got them together and made them case off the buttermilk. They went back to work and soon were making hole right along. Then after a couple of weeks a messenger came in a canoe, bringing an invitation from the Noble, Grand, Worshipful Master, Chancellor, Commander, Exalted Ruler and Jibbonancy for Gib to visit him again. He said he had planned a big party and he hoped Gib would not disappoint him. Now Gib had never been presented at the Court of St. James's as some of his old neighbors back in Vanango County had, but he knew enough to know that an invitation from royalty was the same as a command. He told the crew to stay at work come hell or high water, and then he set out for the royal palace.

He had an even better time than he had had on his first visit, and before he knew it, he had stayed two weeks. When he got back the walking beam was still and the boiler was cold. The men were raising so much hell that he could hear them whooping it up while he was still five miles at sea. They had struck a flow of champagne.

Gib tried to case this off as he had done the buttermilk, but he never could get the crew sober enough to work. In fact he was the only white man on the island sober enough to walk straight, and he couldn't set a casing by himself. Besides, the sight of these drunken brutes was so disgusting that he couldn't stand to look at them. He just got in his launch and went off and left them there, and so far as he knew they were there yet, still drunk as lords.

All of which shows what liquor will do for a man if he fools with it long enough.

Gib cabled for a new rig and a new crew and spudded in on a new location. The drilling went fine. Gib pulled the tools one day when he was down about fifteen hundred feet and found them covered with a whitish liquid.

"There's the buttermilk again," he said to Big Toolie. "Get busy and let's case it off before the other tour finds out about it."

But Gib was tired and thirsty and the buttermilk looked so good that he just had to have a drink of it. He stabbed the bailer into the well and brought up a bucketful. It wasn't buttermilk; it was sweet cream. Still he thought he had better case it off.

While Big Toolie was fixing up the seed bag, Gib had an idea.

"Big Toolie," he asked, "do you suppose these heathens ever heard of ice cream?"

"I don't know," said Big Toolie, "but if they ever tasted it once they never would be able to get enough of it."

"That's what I was thinking," said Gib.

He got the crew together and told them to instal a pump, and he took the first boat for the United States. He bought the biggest and best ice cream factory he could find, pasteurizers, refrigerating machines—everything—and shipped it back with men to operate it. It took every cent he had, but he finally got it up and was ready to begin production. He turned on the steam and started the pump.

Then he discovered that all his fine machinery was useless. While he was gone the cream had soured. And that is the way Gib lost his fortune.

Big Toolie

Among the new crew that had been sent to Gib was one especially fine fellow, a tool dresser twenty-eight inches between the eyes and so tall he could grease the crown pulleys at the top of the derrick without taking a foot from the ground. He did not know what his name was, so he just called him Big Toolie. Big Toolie was a good-natured, playful giant who liked his little tricks. Sometimes Gib would be on the driller's stool, Big Toolie would sneak around and put his foot on the walking beam and either stall the engine or throw the belt off. But he was a good hand and a likeable cuss, and Gib put up with his jokes for a good while. He saw, however, that he would have to find a way to break him. He told him to keep the steam gauge at a hundred and twenty-five pounds. He kept a close watch on him, and one day when Big Toolie was about to step on the walking beam, Gib gave the engine all it had. The beam threw Big Toolie over the derrick and he landed in the slush pit on his head. He crawled out and walked up to Gib and stuck out his hand and said, "Mr. Morgan, I'm your man from now on."

Gib's Wonderful Snake

One time an oil company sent Gib down to South America to dig a test well. They told him to go down ten thousand feet unless he hit oil before he got that far and then to stop. So he had brought only ten thousand feet of cable along. When he got to the end of the cable, however, indications were mighty good and he wanted to go down a little farther. His tool dresser said they would have to shut down a month or two and wait for more rope to come from the States, but Gib didn't like the idea. He hated to delay the work like that, and then besides if the brass hats in the States found out what he was planning to do, they would more than likely stop him. It wouldn't be the first time they had butted in and ruined a good thing.

So Gib said he'd do a little figuring and look around a bit and see what he could do. While Gib was thinking and studying he started walking around, and pretty soon he was out in the jungle. And as he walked along thinking and studying, he came upon a big boa constrictor, a monstrous reptile, twenty blocks long if he was one. The snake had just swallowed a lot of monkeys, Gib figured, for there was a monkey's tail sticking out of his mouth, and he was lying there in a deep stupor not knowing Gib was anywhere about.

Gib went back and got the crew and they dragged the snake back to the rig, spiked his head to a spoke in the bull wheel, and wound him around the shaft, and spliced his tail to the cable. In a few minutes they were cutting ditch again. The snake made as good a drilling cable as you ever saw—a lot better than this new steel cable—with just enough give to make the tools handle easy. Everything went fine for an hour or two. Then the jarring woke the snake up and he began wiggling. The next thing Gib knew he had worked the spike loose and was running off with the whole string of tools. He made for the jungle and got away before Gib could stop him.

Gib hired a bunch of Indians to help him track the snake down.

They trailed him for two weeks and finally found him five hundred miles from the rig. He had dragged the tools around so much that the drill stem was worn down to the size of a crowbar.

Fortunately Gib had brought along another set of tools. He got to thinking about the way he had treated the big snake and he was downright ashamed of himself. He had found him sleeping peacefully and had dragged him off and driven a spike in his head and treated him rough. It was always better to be kind to dumb brutes.

So when Gib brought him back this time, he tried to win his friendship with kindness. First he went to the whiskey well and got a barrel of whiskey and gave it to him. When the snake had swallowed it, he began to wag his tail so friendly-like that Gib felt a deep affection for him. He named him Strickie, and every three weeks he fed him two hundred monkeys. He quit running at night so Strickie could have a rest, and each night when he shut down and unwound Strickie from the bull shaft, he gave him a barrel of whiskey before he put him to bed. Strickie slept in front of Gib's bunkhouse door, and he wouldn't let man or beast come near until Gib ordered him down. It was not long until Strickie was the most valuable piece of equipment Gib had.

After Gib started drilling again and had let out about twenty feet of snake, he pulled the tools to bail out the cuttings. Then he realized how absentminded he had been. The sand line wasn't any longer than the drilling cable. If one wouldn't reach the bottom, the other wouldn't either. So the only thing he knew to do was to unwind Strickie and put him on the sand line. Gib was getting ready to tie his head to the line and his tail to the bucket, but Strickie kept changing ends on him. At first Gib thought he was just being playful. But from the way he kept wagging his tail and sticking it up against the end of the rope Gib saw that he wanted him to tie them together. Gib couldn't see that it made any difference, and so if Strickie preferred to go down head first, he was willing to let him.

As soon as Gib got the splice made, Strickie darted into the

well and began to unwind the sand reel before Gib could stop him
to tie on the bailer. He let him down to the bottom of the well
and reeled him out. Strickie crawled to the sand pit and began to
disgorge pumpings. From that time on Gib never used the bailer.
Strickie had a bigger capacity and could clean the well with one
lowering.

After Gib had been using Strickie about two weeks, the cable
bootjacked off just above the rope socket. Gib let down a horn
socket and caught the tools all right, but as he was trying to pull
them out the latch in the socket broke and he lost them. And
there he was with a string of tools in the well and no way to get
them out.

He had his tool dresser fire up the forge and they went to work
to make a new latch. While they were working, Strickie began
wiggling so that he shook the whole rig. Gib realized that there
wasn't any use in leaving him all wound up with the cable while
they worked maybe the whole day on the horn socket. So he unwound him and gave him his whiskey. As soon as he had swallowed it, Strickie crawled up on the derrick floor to the casing head
and stuck his head in the well and began wagging his tail. Then
Gib caught on to what was in his mind all along. He tied his tail
to the cable and let him down. When he drew him up the tools
came with him. Strickie had swallowed the rope socket, the
sinker, the jars, and half the drill stem. Gib drilled a lot of wells
after that and used a lot of newfangled equipment, but he never
had a better fishing tool than Strickie.

The well went along fine. Gib hadn't found oil, but indications
looked better and better every day. Gib's only worry was that
Strickie might not be long enough to reach the oil he knew was
down there. Each day there would be a little more of Strickie in
the well and a little less on the bull wheels. When he shut down
one Saturday night there was just enough of Strickie left to wrap
once around the shaft. Gib had got a showing of oil that day,
and he knew that he was right on top of something big and he

stayed awake all that night worrying about just how to get to it.

But the next morning all his worries were over. That very night Strickie shed his skin and Gib had plenty of cable to finish the well.

It was a gusher, but Gib closed it in and soon had it running into the storage. In no time at all it had filled the one thirty-thousand barrel storage tank. Gib shut it in and began laying a gathering line to the coast, where the big new tanker steamers that Standard had just put on could load. His pipe ran out when he was still about half a mile from the harbor, but by that time Strickie had shed again and his skin not only finished out the line but left enough over to reach up on deck and fill the compartments without all the bother of making pipe connections.

Gib's Hardest Fishing Job

In all his years of drilling Gib never had a lawsuit. But he came mighty near it one time. He was working for a company in Ohio. It wasn't a very big proposition, and as it turned out he was glad that it wasn't. Most of the acreage was leased before his company got in there, and all they could get was a hundred-and-sixty-acre farm, a wedge-shaped piece of land in Washington county, at least two miles from production.

At the depth of a thousand feet Gib struck a vein of granite that would take the edge off of a bit in no time at all. He spent more time pulling tools and dressing bits than he did in drilling. Even then he couldn't keep a flare on the bit. It would stick in the hole; he would jar it loose and dress it, and it would stick again. This went on from twelve to four o'clock in the afternoon, when the tools lodged so tight he couldn't budge them an inch. He knew that he was in for a fishing job that was a real fishing job.

He ran his knife down the sand line and cut the cable a few feet above the rope socket. Then he rigged up the longest string of fishing tools ever let down in an oil well. On the bottom was a rope spear to catch the cable end. Above that was a sinker, and above that a pair of jars. He put on so many sinkers and jars that he couldn't keep count of them and had to send for a bookkeeper. Then came the drill stem and rope socket.

Gib lowered the string of tools into the well and right away he caught the tools. Then began the job of jarring them loose. He worked all the rest of the afternoon, but when dark came he couldn't tell that he'd moved them an inch. At ten o'clock that night they were still as tight as ever. At midnight they were still stuck, but Gib and his toolie were glad to see twelve o'clock come so they could go home and sleep and let the other tour do the fishing for a while. But the other tour didn't come. At twelve-thirty they still hadn't showed up. The tool dresser wanted to quit and go to bed, but Gib told him they'd better work on a while longer. He'd see that the other men made it up to them sometime.

They worked until two o'clock and still no relief came. Gib told the toolie to watch her a while and he would walk back to the boardinghouse and see what was the matter. Gib hadn't gone far when he found himself on the edge of a precipice. It was pretty dark that night and he was lucky that he didn't walk right off of it. Down below he saw a lantern and heard hammering and sawing.

"What are you doing down there?" he yelled.

"Hello, Gib," the other driller answered. "We're building a ladder. We're trying to get up to the well."

Gib had jarred the whole lease up seventy-five feet. The grave-yard tour had come out at twelve as they were supposed to, and had found the precipice and had begun work on the ladder right away. But as Gib kept jarring the lease higher, they had to go back several times for more lumber, and even then the lease was gaining on them.

Gib shut down and they finished the ladder. It was daylight and they were still wondering what to do when the farmer who owned the land came out and said he was going to have the law on him for ruining his farm. He had gathered a load of corn the day before and was ready to take it to market and his wife was going with him, and they had got up before daylight to get an early start, and there they were stuck up seventy-five feet in the air and no way to get down, and the farm would never be worth anything again. Now Gib knew that the lease had a clause in it about damage to crops and premises, and he was afraid that the company was in for a lawsuit. However, he gave the man a drink out of a jug he kept around the rig for emergencies and told him that if he would give him a little time, he thought he could fix it up.

So Gib began thinking and studying and before long he hit upon a solution so simple that he wondered why he hadn't thought of it sooner. He sent to a big medicine house and got forty barrels of arnica salve, hoisted them up to the rig on the sand line, and packed the salve into the well. In an hour and a half the lease had fallen two feet. By the end of the second day the swelling was completely gone and the rock was so softened that the tools came out with no more trouble than bealing out a splinter.

Since Gib Morgan was known as the Munchausen of the oil fields at least as early as 1880, he was obviously the first of the tall tale heroes of the industry. His displacement by Paul Bunyan must have occurred between 1901 and 1920. The displacement, however, was not complete. The men who knew Gib Morgan in the older oil states and their children still cherish his tales and tell them with an enthusiasm and in a detail not found in the Bunyan versions. Bunyan, for example, gets his tools stuck, puts on fifteen sets of jars, and "the terrific pounding jarred the whole lease fifty feet above the surrounding land before he broke the rock." The tellers of the Morgan version make use of suspense— Gib keeps wondering why relief doesn't come—and never omit the building of the ladder.

But the tall tale hero, whether he is called Morgan or Bunyan, belongs to the past. Modern roads and road-building equipment have ended the isolation of the oil field worker. Modern drilling and ditching equipment have changed him from a back and

muscle man to a skilled technician. He has lost interest in the tall
tale. Creativity has ceased.

The relation of the tall tale hero to the people has not been sat-
isfactorily defined. The traditional belief, accepted perhaps by a
majority of folklorists, is that the hero is in some rather direct and
detailed way a projection of the folk, or such of the folk who enjoy
telling or hearing the tales. A corollary is that the tall tale is created
to relieve man's fears, the hero's victories over nature and man
and beast representing responses to a hostile environment.

There is little evidence to support this theory in relation to
Anglo-Saxon America. Before the opening of the nineteenth century the Enlightenment had driven the demons from the woods,
and the Pennsylvania rifle had given the American confidence to
cope with his natural enemies. He knew that the Indian was a
formidable adversary, but he knew that he could be killed by the
same means other men could. He knew that the panther might
turn on him when cornered, but he knew that ordinarily he was
afraid of men. Frontier stories of ferocious animals were mostly
for the benefit of tenderfeet. The irrational fear that persisted longest was the fear of ghosts; yet no ghost-slayer arose among the
folk heroes.

One valid observation, I believe, is that a literature of exag-
geration, whether oral or written, flourishes best in a climate of
excitement, of optimistic expectancy, which is not the climate of
today. One such period was the Renaissance in Europe. In America one such period was the Jacksonian era, an era of industrial
development as well as westward expansion. Crockett, symbol
of the backwoods, emerges as the leading folk hero.

In the oil industry there have been two such periods. One followed the expansion of the industry from Titusville into other
Appalachian areas. The excitement was not altogether local. A
new resource had been utilized, and nobody knew its limits. This
was the age of Gib Morgan. The other followed the coming in
of the Lucas gusher at Spindletop on January 10, 1901, the most

revolutionary event in the history of oil since Drake's well. News of it went all over the western world. Thousands flocked to Beaumont to see it. It indicated that there was plenty of oil, not only for the decreasing number of lamps, but for the fueling of locomotives, ships, and the new horseless carriages. For a brief time Paul Bunyan flourished.

Much remains to be explained about the workings of the imagination, but the ingredients that go into works of the imagination can often be identified. It is not difficult to find in reality the stimulus that led to the creation of a tall tale. There was, for instance, in Pennsylvania, a formation known to workers as the buttermilk sand, since the cuttings in the slush pit had the color and consistency of buttermilk. So the striking of buttermilk was a commonplace experience. But if buttermilk, why not other things associated with it, such as cornbread and turnip greens? Or sweet cream or bay rum or whiskey, or any liquid? The whiskey well of Oklahoma came after Gib Morgan's. If an article in the *New York Herald* is to be trusted, a driller did strike whiskey in 1898. A well was being drilled on top of a hill. At about two hundred feet "the drill seemed to fall into a cavity" and fumes of whiskey came to the surface. The driller drew out the tools, emptied his lunch bucket, and lowered it on the bailer line. It came up, he said, full of "the best corn whiskey I ever laid my lip to." The explanation was that he had drilled into a moonshiner's cave.

The tale of the perpetual motion rig, whether the tools are set in motion by striking a vein of rubber or the root of a rubber tree or whether the tools themselves are made of rubber, probably had its origin in the self-drilling wells reported from time to time. One such well was brought in on Jacob Miller's farm in Pennsylvania in 1870. Before the tubing could be set the derrick burned to the ground. "The tools are still in the well," reported the Oil City *Times* on May 3, "and the occasional flow of gas raises them a number of feet, when they fall back with great force, each time increasing the depth of the hole. By this process the well has drilled itself upward of fifty feet within the past two weeks." Before the tools were recovered in June, they had gone another fifty feet. Another example was the Seneca Oil Company's well near Bradford. When the well came in on May 27, 1898, the oil was ignited by the boiler fire, and the crew ran for their lives. The next day the well drilled itself through the oil sand and struck a flow of salt water, which put out the fire. The water was cased off and the well produced oil.

Some tales are fanciful solutions of technical problems. The big derrick story as Gib Morgan told it is an example. In some formations it was impossible to drill with a cable tool rig. Here is a veteran driller's explanation of the failure of the equipment at Spindletop. They would go through a rock and hit quicksand. Then they would hit another rock and more quicksand, and soon the hole would be so small they couldn't get their tools in it. One solution would be to start with the biggest substitute for casing available, that is steel storage tanks, which would be handled only with an immense derrick. As the technical problem is solved, this one by rotary drilling and the use of mud, the relation to reality is obscured and the tale loses point.

15

Anecdotes

The Boom

A MAN ARRIVED at an oil town at the peak of the boom. There were no rooms at the hotel, not even a cot in the hall. He spent the night on a bench in the railroad station and early in the morning renewed his search for a room. He spent the day going from house to house, but at nightfall had not found a place to sleep. He walked into a tent at which he had applied earlier to ask if a cot had become available. The proprietor said that one of his men had just got killed on the rig.

"Could I have his cot?"

"I reckon so if you got three dollars a night, but you'll have to get out in eight hours because it's rented to two other men."

"Isn't three dollars rather high for a cot for eight hours?"

"Take it or leave it."

"I suppose you'll put on a clean sheet."

"You ain't no better than the other men that sleep on this cot. I figure that a sheet that's good enough for them is good enough for you."

"How about a pillow?"

"That will be two bits extra."

"Well, I guess I'll have to take it."

The proprietor left and a moment later the guest heard a shot. He went out and found the proprietor.

"Did you hear a shot?"

"Yes, I heard it."

"What happened?"

"I had to shoot a man."

195

"You shot a man! What for?"

"He kicked about my accommodations."

An oilman, hearing that oil had been struck at a certain place, rushed there as fast as he could. The place was isolated and the way was long, and when he got there he could not find a place to sleep. After an unpleasantly cool night on the ground, he began asking everyone he met if he had heard about the big oil strike in the Mojave Desert. The question was repeated and soon became a general rumor. Details were added about the depth, the yield, and the quality of the crude. By nightfall there were plenty of rooms available. The oilman, however, did not take one. He said that if that many people believed a rumor, there must be something to it. So he followed the crowd.

A roughneck died and found himself at the gate of Heaven. But he couldn't get in because there was a big mob of other rough-necks waiting for Saint Peter to admit them.

So he started a rumor.

"Say, buddy," he told one of his fellow-workers. "I hear there's a big boom on in Hell. Bringing in lots of wildcats and paying big money."

The rumor spread, and had its effect. All the oil workers turned and headed for the big boom in Hell.

As he stood there watching them run down the road, the old roughneck said to Saint Peter:

"You know, the way they are taking off—there might be something to it."

And he ran off down the road after them.

And they say that's the only reason there are any roughnecks in Hell today.[1]

The Boardinghouse

A wildcat well was being put down on a farm in the apple

country of Arkansas. The drilling crew was boarding at the house of the farmer. Each morning there would be fried apples for breakfast. Each noon there would be stewed apples for dinner. Each night there would be apple cobbler for supper. This diet was welcomed at first, but as the weeks went by the men ate with less and less gusto. One day as the apple dish went by, a driller passed it on. The landlady said, "Mr. Green, I'm afraid you don't like apples."

"O, yes," he said. "I like apples. I'm just not a son of a bitch about them."[2]

Once in Ranger a driller named Grant Emory was staying at a boardinghouse where meals were served family style. Bowls and platters of food would be placed on the table and the boarders would help themselves. Usually there would be hot biscuits for breakfast. One morning Emory broke open a biscuit and put in an enormous hunk of butter. The landlady, who was standing by, said, "Mr. Emory, do you know that butter is costing me seventy-five cents a pound?"

Emory cut off another hunk of butter as big as the first and said, "And worth every damn cent of it, too."[3]

At a boardinghouse for oil field workers meals were announced by the ringing of a bell. The graveyard shift (the crew working from twelve midnight to twelve noon) had come in and were sitting on the porch waiting for the bell. Nearby asleep in the shade of a shrub was a long-eared hound. The bell woke him up and he let out a long loud howl. A driller looked at him contemptuously and said, "What are you howling about? You don't have to eat it."

One Ranger

A Ranger was sent to an oil town where the lawless element had gotten out of hand. Some miles from the town he mounted a horse he had previously arranged for and rode slowly along the side of the highway. Passing motorists recognized him for what he was, as he intended them to, and long before he reached town he was expected.

He had many years before seen Buffalo Bill's Wild West Show, and had not forgotten the act in which Bill tossed glass balls in the air and broke them with rifle fire. Later he had heard that Cody had loaded the shells with shot. He so loaded three of the five shells he carried in his six-shooter. At the edge of town he bought an apple, upon which he munched as he rode into the section where the lawless element hung out. There he threw the core in the air and shot it three times before it hit the ground. An astonished onlooker asked if he might see the gun. The Ranger handed it to him and he saw that the remaining shells were loaded with bullets. Many of the tough characters left town in a hurry and none resisted arrest.

Two men were walking down the street of an oil town to which a Ranger had been sent. As they passed a building they heard a shriek and a groan followed by other shrieks and groans. One of the men walked on as if he had heard nothing. The other asked, "Did you hear that?"

"Yes," said the other, as he continued walking, "I heard it."

"Somebody's in trouble."

"Yes, I know it."

"Don't you think we ought to go in and see about it?"

"No, it's none of our business."

"But somebody's being hurt. We ought at least to call the police."

"O come on. It's only the Ranger taking a voluntary statement."

The Bulldog

Sometimes during drilling operations a piece of equipment may be lost in the well. It is then called a fish, the job of recovering it is called a fishing job, and the devices used are called fishing tools. If the tool fits inside the fish, it is a spear, and a spear with an automatic latch that cannot be released until it is brought to surface is called a bulldog spear, often shortened to bulldog.

This was the tool that a drilling superintendent working for an American oil company in Mexico in 1907 needed to complete his set. An engineer whose previous experience had been in railroading had recently been assigned to the field. After a brief period of inspection he was preparing to return to the States for a short visit before taking up a long residence in Mexico. He told the field superintendent that if there was likely to be need for any equipment not available in Mexico, he would attend to its purchase while in the United States. The superintendent replied that he might have some fishing to do and that he would like to have a sixteen-inch bulldog. When the engineer returned the superintendent met him in the hotel lobby, and after some conversation asked, "Did you get me a bulldog?" "Yes," said the engineer, "he's in my room. I'll get him right now." He returned leading a brindle bulldog approximately sixteen inches long.[4]

Rotary Club

This story is usually set in Mexia, which is where it should have happened, if it had happened at all. The Mexia boom came just as the Ranger boom was ending, and drillers from the Ranger area rushed to Mexia looking for work. There they were disappointed, for they were cable-tool men and did not know how to operate the rotary equipment being used in the new field. This meant that if they found work at all, they had to begin as green hands at a considerable loss of pay and prestige.

One day some cable-tool drillers were walking down the street when one noticed a Rotary Club symbol on a restaurant window. "Look," he exclaimed to the others. "The sons of a bitches even have a club!"

Tales of Hoffman

Pete Hoffman began as an oil field worker in West Virginia, became a driller, then a drilling contractor, and eventually an independent operator. He followed oil production into the Southwest, and died in Texas, a moderately wealthy man.

He became a legendary character whose fame rested chiefly upon his skill as a borrower. For many years drillers, drilling contractors, and small operators borrowed freely from each other. If X needed a bulldog spear or an underreamer or a similar piece of equipment, he might to save either time or money, go to the nearest rig and make his need known. If Y had the equipment and was not using it, he would lend it freely and borrow just as freely if he had need to.

Borrowing was usually confined to relatively small articles not essential for ordinary operations. According to tradition Pete Hoffman was the first, though not the last, to borrow a boiler with a full head of steam. That was back in West Virginia at a time when the standard boiler was mounted on wheels and was small enough to be drawn by a pair of horses. The crew had shut down the rig and gone to lunch. When they returned the boiler was gone. They went to Hoffman's rig and found it connected to his engine and in use. He said he had borrowed it for the afternoon and would return it as soon as he finished setting casing. He kept his word.

At about this time Carl Angstadt was in charge of the Jacuzzi supply house at Sistersville. One day he missed a boiler from the material yard. He thought perhaps a clerk had sold it and had forgotten to make the charge, but nobody could recall selling it. He went over his records for several years and listed every customer who had bought a boiler. Assuming that nobody would pay for the missing boiler unless he had it, he billed each one. On the first of the month he got a check from Pete Hoffman.[5]

In view of the widespread custom of borrowing it was almost

impossible to secure a conviction for theft if the accused was an oilman who had use for the property and did not offer it for sale. The defense would be that he intended to return it. This was not Pete Hoffman's defense, however, on the one occasion he was brought into court. He was accused of stealing an eight-inch cable-tool bit, and the bit was in the courtroom as exhibit A. The facts implicating Hoffman were admitted by both sides. The bit, whether the one stolen or one identical with it, had been found in his possession. Fresh wagon tracks had been seen on a road some yards from the rig, but they did not lead to it. The wagon had not left the road. Leading from the road to the derrick and back was one and only one set of footprints. Somebody had left the wagon, walked to the well, and returned. Hoffman's defense was that no man could have carried the bit, which weighed over four hundred pounds, from the rig and loaded it on the wagon. The jury was convinced and acquitted him. He then asked the judge who the bit belonged to. "Is it mine?" The judge supposed so, since he had been acquitted of stealing it. Hoffman picked up the bit and carried it out of the courtroom. Nobody seems to know just where this happened. It might not have happened at all. Tales attached themselves to Pete Hoffman.[6]

East Texas Episode

The standard oil lease is for ten years or until the drilling of a well, and an annual rental is paid the lessor. Usually some bank is designated as the depository agent and deposit of the rental money in the bank constitutes legal payment. In East Texas during the depression days of the 1930's some of the banks so designated failed, and the oilmen in order to retain their leases found it necessary to tender lease money in person. One day a representative of an oil company arrived in the Sulphur Bottom country with $120 in currency for a man named Tobe Fails, who owned a rundown farm in the region. He stopped at a filling station and asked how to find the Fails's farm. The attendant pointed to a rusty and topless Model T Ford and said, "That's Tobe turning off the highway now."

The agent overtook Fails on the dirt road, but when he sounded his horn the old car leaped ahead and began dodging through the trees and around piles of bush where tie cutters had been at work. After a long and circuitous race the model T stopped back of a house. The agent stopped in front, got out, and knocked on the door. After a few minutes Tobe opened the door part way and looked out. "What do you want?" he asked.

The oilman held up some bills. "I want to tender you the hundred and twenty dollars rental due you on that oil lease my company bought last year."

"Well, of all the luck!" Tobe said. "You have been chasing me all over hell and half of Texas to pay me a hundred and twenty dollars, and I just throwed two gallons of the best whiskey I ever made in the well."[7]

Ellen Burger

A well was going down on a three-hundred-acre farm in Coleman County, Texas. The farmer had heard geologists and scouts talk about the possibilities of oil. There were several producing sands in the region, often referred to as the Ranger, the Caddo, the Palo Pinto, the Morris, the Strawn, and the Ellenberger formations. Of these the deepest was the Ellenberger. The well went down without encountering any of the shallow formations and now only the Ellenberger was left. The drill was nearing the depth at which it would be struck, if it was to be struck at all.

The farmer spent nearly all his time at the well, remaining at night as long as he could stay awake. One morning he staggered to bed at two o'clock. But he did not get much rest. He rolled and tossed and talked in his sleep, saying over and over, "Come through, Ellenberger, old girl. Don't let me down. Come through."

At daylight he heard his wife in the kitchen. He got up and

went in. There was no smell of coffee or bacon. He asked her if she had any late news from the well.

"No," she said, "I haven't heard anything from the well, but I heard you talking in your sleep. Who is this Ellen Burger? You better tell her to cook your breakfast."[8]

Texas Oil Millionaire

A newly-made Texas oil millionaire held open house upon the completion of his ranch house mansion. What he was proudest of was his swimming pools. As he pointed out to his guests, there were three; one filled with chilled water for those who wanted a good cooling off, one filled with warm water for those who preferred a gentle swim, and one with no water at all for those who did not wish to swim.

A hitchhiker held up his thumb. A big Cadillac with a Texas license came to a stop. The door opened, the hiker got in, and soon the car was speeding down the highway, weaving in and out of traffic. Then the hiker noticed a pair of thick-lensed glasses on the seat. He looked at the glasses and then at the driver.

"Don't worry, son," the driver said; "I can see just fine. You see, I got the windshield ground to my prescription."

An oil magnate went to the courthouse and asked if he might get an automobile driver's license for his son.

"How old is your son?" asked the clerk.

"He's ten."

"We can't give him a license."

"He's going to be terribly disappointed. I just bought him a new Cadillac."

"You ought to know that ten-year-old boy shouldn't be trusted on the streets with a car."

"O, he won't be on the streets. He just wants to drive it around the house."

A Texas millionaire went into a church to pray. He knelt and said:

"O Lord, I know everything has to operate according to your will. But I understand you listen to prayers, so here I am. First, Lord, I want to tell you about my ranch out in the hill country. It's not a great big spread, just three thousand cows in my foundation herd, good white-faced stuff. It's getting pretty dry up there, and if it don't rain in a couple of weeks, I'll have to start feeding. So if you could send a rain that way, I sure would appreciate it.

"Then, Lord, I bought some land up in Colorado. Supposed to be uranium land, but I don't know how it'll turn out. But if you could let it turn out good, I sure would be grateful.

"Then, Lord, I've got sixteen oil wells drilling. Of course, Lord, everybody knows that a man can't expect to get oil in every wildcat he drills, and I'm not asking for oil in all sixteen. But, Lord, if you could let me have oil in twelve, I'd sure be obliged to you. Amen."

While the oilman was praying, a man in a threadbare suit had entered quietly and taken a seat. He now knelt to pray. He said, "O, Lord, I've done everything I know to do, and I don't know anything else to do but pray. I'm an accountant, and I lost my job when the firm sold out to another company. That was six months ago and my savings are all gone. It's hard for a man forty-seven years old to find a new job, but I've tried. I've worn out my shoes going from one place to another. Then my suit is beginning to look shabby, and I have a wife in the hospital. The doctor says she can go home, but I owe the hospital a hundred and seventy dollars and they won't let her go. And every day they add twelve dollars to the bill. O Lord, please help me in some way to get on my feet again. Amen."

The oilman pulled out his wallet and peeled off five one-hundred-dollar bills and quietly placed them in the accountant's hand. Then he knelt again and said, "Lord, you needn't bother about this little chicken—I've taken care of him."

Notes

CHAPTER 1

1. Stanley Vestal, *Short Grass Country* (New York, 1941), p. 289.
2. Wallace Pratt, *Oil in the Earth* (Lawrence, Kansas, 1942), p. 8.
3. Kenneth W. Porter to M. C. B., July 27, 1949.
4. *Palestine Times,* March 24, 1927.
5. Don Powers, interview.
6. Boyce House, *Were You in Ranger?* (Dallas, 1935), pp. 13-23.
7. *Ibid.*
8. Carl Coke Rister, *Oil! Titan of the Southwest* (Norman, 1949), pp. 150-51.
9. J. W. Turner, interview; letter to M. C. B., April 9, 1945. In the suit Mr. Turner was attorney for R. E. Barker *et al.*
10. *New York Times,* November 14, 1934.
11. *New York Times,* February 5, 1937.
12. R. S. Kennedy, tape-recorded interview, July 30, 1952.
13. *New York Times,* July 5, 1931.
14. *Oil Investors' Journal,* XII (February 10, 1910), 23.

CHAPTER 2

1. Mrs. Fowler's dream has been widely publicized. For a typical statement, see the *Wichita* [Falls] *Times,* October 11, 1936. Boyce House in his account of the Fowler discovery well *(Oil Boom,* Caldwell, Idaho, 1941) does not mention the dream, and Walter Cline, one of the organizers of the Fowler Farm Oil Company and drilling contractor for the well, does not vouch for it.
2. The incident of the Indiana salesman is from Samuel Tait, Jr., *The Wildcatters* (Princeton, 1946), p. 74.
3. The story of the Coquette well is told in J. H. A. Bone, *Petroleum and Petroleum Wells* (Philadelphia, 1865), pp. 13 ff., and in Herbert Asbury, *The Golden Flood* (New York, 1942), pp. 109 ff.
4. Mrs. Weger's dream was told to me by John C. Chambers in an interview at Chanute, Kansas, April 25, 1942.
5. The Rust story was obtained from John Rust in a tape-recorded interview at Borger, Texas, September 9, 1952.

CHAPTER 3

1. J. H. A. Bone, *Petroleum and Petroleum Wells* (Philadelphia, 1865), p. 35.
2. *Austin* (Texas) *Statesman,* October 8, 1947.
3. Riley Maxwell, interview, January 2, 1942.

4. H. H. Adams, interview, April 4, 1942.
5. Guy Findley, tape-recorded interview, May 5, 1956.
6. *New York Times,* October 8, 1922, II, 5:1.
7. Evelyn M. Penrose, "Dowsing," *Blackwoods Magazine,* CCXXXII (September, 1932), 345-53.
8. Adams, interview.
9. Adams, interview.

CHAPTER 4

1. Herbert Asbury, *The Golden Flood* (New York, 1942), pp. 269-71.
2. Dorsey Hager, "Lady Luck," *Oil and Gas Journal,* XXXIII (January 3, 1935), 50.
3. *Ibid.*
4. George W. Weller, tape-recorded interview, July 22, 1953.
5. Ruth Bryant, tape-recorded interview, August 2, 1959.
6. Annie Webb Buchanan Jackson, tape-recorded interview, April 6, 1956; William A. Owens, "Seer of Corsicana," *Southwest Review,* XLIII (Spring, 1958), 124-34.

CHAPTER 5

1. J. Frank Dobie, *Coronado's Children* (Dallas, 1934), pp. 74-78.
2. A. C. White, "An Interview with a Doodlebug Expert" (unpublished MS).
3. H. B. Duetsch, "Taming The Wildcat," *Saturday Evening Post,* CCX (June 11, 1938), 14 ff.
4. L. W. Blau, "Black Magic in Geophysical Prospecting," *Bulletin of the Society of Petroleum Geophysicists,* I (January, 1936), 1-8; Lois Osburn, "Oil Finders" (unpublished MS).
5. This device was described and stock in Radioscope Laboratories Company was offered in advertisements in the *Fort Worth Star-Telegram* for November 20 and 27, 1921.
6. M. G. Cheney, interview, April 4, 1942.
7. H. H. Adams, interview, April 4, 1942; "Doodlebugs and Doodlebuggers," *California Folklore Society Quarterly,* III (January, 1944), 53-58.
8. W. M. Hudson, Sr., tape-recorded interview, September 18, 1952.
9. Allen Hammill, tape-recorded interview, September 2, 1952.
10. Benjamin Coyle, tape-recorded interview, July 30, 1953.
11. O. W. Killum, tape-recorded interview, September 5, 1956.

CHAPTER 6

The story of Dorset and his daughter I obtained from John C. Chambers in Chanute, Kansas, April 25, 1942.

Information about Jackson Barnett may be found in E. P. Atwood, "The Crazy Snake Rebellion," *Vassar Undergraduate Studies* (May, 1942); C. B. Glasscock, *Then Oil Came* (Indianapolis, 1938); and Gerald Forbes, *Flush Production* (Norman, 1942). The case was widely reported in the newspapers from 1917 to 1945. The *New York Times* has been especially useful.

The Osage Indian affairs are discussed in Grant Foreman, *A History of Oklahoma*

(Norman, 1942), and in the Glasscock and Forbes books cited. I have also consulted pertinent statutes and treaties. For the financial statement I am indebted to H. T. Ferrier, Chief, Branch of Minerals, Osage Agency, Pawhuska, Oklahoma.

Information concerning the University of Texas comes from H. Y. Benedict, *A Source Book Relating to the History of The University of Texas* (University of Texas Bulletin No. 1757 [Austin, 1917]), from the office of the auditor of the university, and from the annual fiscal reports of the university.

CHAPTER 7

1. Norman Cousins, "Mr. D. and His World," *Petroleum Today* (Spring, 1960), p. 12.

2. Ruth Sheldon Knowles, *The Greatest Gamblers* (New York, 1959), p. 300.

3. W. C. Gilbert, tape-recorded interview, July 22, 1953.

4. Boyce House, *Were You in Ranger?* (Dallas, 1935), p. 81.

5. Dorsey Hager, "Lady Luck," *Oil and Gas Journal*, XXXIII (January 3, 1935), 50.

6. *Ibid.*, p. 51.

7. James A. Clark and Michel T. Halbouty, *Spindletop* (New York, 1952), p. 42.

8. Hager, *op. cit.*

9. Hager, *op. cit.;* Knowles, *op. cit.*, p. 230.

10. J. H. Plenn, *Saddle in the Sky* (New York, 1940), p. 206.

11. Samuel W. Tait, Jr., *The Wildcatters* (Princeton, 1946), p. 95.

12. Hager, *op cit.*

13. W. C. Gilbert, tape-recorded interview, July 22, 1953.

14. O. W. Killum, tape-recorded interview, September 15, 1956.

15. C. A. Warner, *Texas Oil and Gas Since 1543* (Houston, 1939), p. 59.

16. H. A. Rathke, tape-recorded interview, September 13, 1953.

17. Tait, *op. cit.*, p. 113.

18. George Sessions Perry, *Texas: A World in Itself* (New York, 1942), pp. 173-75.

19. J. R. Webb, tape-recorded interview, August 1, 1952.

20. Boyce House, *Oil Boom* (Caldwell, Idaho, 1941), p. 39.

21. *Ibid.*, p. 40.

22. Isaac Marcosson, "Texas Oil Domain," *Saturday Evening Post*, CLCVI (April 19, 1924), 16.

23. George W. Gray, "The Roaring Tides of the Oil Fields," *New York Times* (September 20, 1931), v, p. 40.

24. Walter Cline, tape-recorded interview, August 13, 1952.

25. Quoted by Martin Schwettman, "The Discovery and Early Development of the Big Lake Oil Field" (Master's thesis, University of Texas, 1941), p. 101.

26. *Ibid.*

27. Hager, *op. cit.*, p. 51.

28. *Ibid.*

29. House, *Were You in Ranger?*, pp. 9-10.

30. Quoted by Schwettman, *op. cit.,* pp. 101-2.

31. A. I. Levorsen, *Geology of Petroleum* (San Francisco, 1956), p. 638; Knowles, *op. cit.,* p. 301; *Bulletin of the American Association of Petroleum Geologists,* XXX (June, 1946), pp. 813 ff.

CHAPTER 8

1. Samuel W. Tait Jr., *The Wildcatters* (Princeton, 1946), p. 148. More recent writers, for example Carl Coke Rister, *Oil! Titan of the Southwest* (Norman, 1949) and Ruth Sheldon Knowles, *The Greatest Gamblers* (New York, 1959), do give considerable attention to geology.

2. M. G. Cheney, interview, April 4, 1942.

3. Charles N. Gould, *Covered Wagon Geologist* (Norman, 1959), p. 118.

4. A. I. Levorsen, *Geology of Petroleum* (San Francisco, 1956), p. 140; Knowles, *op. cit.,* pp. 72, 73.

5. Allen Hammill, tape-recorded interview, September 2, 1952.

6. Ed Prather, tape-recorded interview, April 4, 1954.

7. Benjamin Coyle, tape-recorded interviews, July 28 and 30, 1953.

8. Gould, *op. cit.,* p. 178.

9. Lew Allen, "Oil Wildcatters," *New York Times* (July 30, 1922), vii, 3.

10. Boyce House, *Oil Boom* (Caldwell, Idaho, 1941), p. 141.

11. Sidney Paige, tape-recorded interview, June 7, 1954.

12. W. H. Bryant, tape-recorded interview, July 29, 1952.

13. Levorsen, *op. cit.,* p. 14.

14. On the early history of petroleum geology see J. V. Howell, "Historical Development of the Structural Theory of Accumulation of Oil and Gas," in *Problems of Petroleum Geology,* Sidney Powers Memorial Volume, edited by W. E. Wrather and F. H. Lahee, American Association of Petroleum Geologists (Tulsa, 1934), pp. 1-23; Ralph Arnold, "Two Decades of Petroleum Geology, 1902-1922," *Bulletin of the American Association of Petroleum Geologists,* VII (1923); Levorsen, *op. cit.,* pp. 138-41; Gould, *op. cit.;* Knowles, *op. cit.*

15. Levorsen, *op. cit.,* p. 140; Knowles, *op. cit.,* p. 73.

16. E. I. Thompson, tape-recorded interview, September 3, 1952.

17. Knowles, *op. cit.,* p. 153.

18. J. T. Young, tape-recorded interview, August 15, 1952.

19. Knowles, *op. cit.,* pp. 181-82.

20. C. A. Warner, *Texas Oil and Gas Since 1543* (Houston, 1939), pp. 355, 375.

21. The most thorough account is that in James A. Clark and Michel Halbouty, *Spindletop* (New York, 1952).

22. Warner, *op. cit.,* p. 19.

23. Patillo Higgins, tape-recorded interview, July 25, 1952.

24. Kenneth W. Porter to M. C. B., July 27, 1949.

25. Clark and Halbouty, *op. cit.,* pp. 21-22.

26. Clark and Halbouty, *op. cit.,* p. 24-26; House, *op. cit.,* p. 23.

27. Knowles, *op. cit.,* p. 28; Clark and Halbouty, *op. cit.,* pp. 36-37.

28. Knowles, *op. cit.,* p. 29; Clark and Halbouty, *op. cit.,* pp. 27-38; Rister, *op. cit.,* p. 53; Warner, *op. cit.,* p. 35.

29. Henrietta M. Larson and Kenneth W. Porter, *History of the Humble Oil and Refining Company* (New York, 1959), p. 16.

30. Stanley Walker, "Where Are They Now? Mr. Davis and His Millions," *New Yorker*, XXV (November 26, 1949), 35-47.

31. *Ibid.*

32. David Donoghue, interview notes, 1947; Rister, *op. cit.*, pp. 174-77.

33. Lockhart *Post-Register*, August 3, 1936.

34. Henderson *Daily News*, November 30, 1938; Dorman H. Winfrey, "A History of Rusk County" (Master's thesis, University of Texas, 1951), p. 145.

35. Hattie Roach, *A History of Cherokee County* (Dallas, 1934), p. 80.

36. Rister, *op. cit.*, p. 366; Knowles, *op. cit.*, p. 267.

37. Knowles, *op. cit.*, p. 268; Larson and Porter, *op. cit.*, p. 397.

38. Rister, *op. cit.*, pp. 327-29; Knowles, *op. cit.*, p. 252; Harry Harter, *East Texas Oil Parade* (San Antonio, 1934), p. 41; Henderson *Daily News*, October 3, 1930.

39. Henderson *Daily Times*, September 23, 1930.

40. Arnold, *op. cit.*, p. 614.

41. Knowles, *op. cit.*, p. 146; Rister, *op. cit.*, p. 193.

42. Knowles, *op. cit.*, p. 240.

43. *Ibid.*, p. 280.

CHAPTER 9

1. See "Trickster," *Standard Dictionary of Folklore, Mythology and Legend* (New York, 1950).

2. *Oil on the Brain Songster* (Cincinnati, 1865), p. 19.

3. Both quoted in *The Derrick Handbook of Petroleum* (Oil City, Pennsylvania, 1898), I, 1047-49.

4. Alfred Wildon Smiley, *A Few Scraps (Oily and Otherwise)* (Oil City, 1907), pp. 72-74.

5. John Wynne, tape-recorded interview, January 26, 1960.

6. E. M. Everton, oral interview, November 20, 1941.

7. James A. Haakerson, oral interview.

8. *Were You in Ranger?* (Dallas, 1935), p. 116.

9. Boyce House, *Oil Boom* (Caldwell, Idaho, 1941), p. 32.

10. *New York Times*, May 9, 1923.

11. *Oil Investors' Journal*, II (June 1, 1903), 5.

12. J. K. Barnes, "Doctor Cook's Discovery of Oil," *World's Work*, XLV (April, 1923), 611-17.

13. "Investing in Oil," *Review of Reviews*, LXIII (April, 1921), 443.

14. *Oil Investors' Journal*, III (October 15, 1904), 7.

15. House, *Oil Boom*, p. 26.

16. *Oil Investors' Journal*, III (May 3, 1905), 9.

17. Barnes, *op. cit.*

18. *Current Opinion*, LXX (April 21, 1921), 545.

19. *Literary Digest*, LXXVII (December 8, 1923), 11.

20. *Baptist Standard*, May 2, 1901.

21. "Investing in Oil," *op. cit.*, pp. 443-44.

22. *New York Times,* March 9, 1924.

23. *Fort Worth Star-Telegram,* January 15, 1922.

24. House, *Oil Boom,* pp. 56-57.

25. Barnes, *op. cit.,* pp. 614-15.

26. House, *Oil Boom,* p. 31.

27. James A. Clark and Michel T. Halbouty, *Spindletop* (New York, 1952), p. 87.

28. House, *Oil Boom,* p. 31.

29. Carl Coke Rister, *Oil! Titan of the Southwest* (Norman, 1949), pp. 185-88.

30. Wirt Hord, *Lost Dollars, or Pirates of Promotion* (Cincinnati, 1924), p. 13.

31. Rister, *op. cit.,* p. 394.

32. *World's Work,* XXVII (November, 1918), 510.

33. *New York Times,* April 7, 1923.

CHAPTER 10

1. Thomas P. Smiley, "Shooter's Life Loses Old-Time Thrills," *Oil and Gas Journal,* XXVI (February 10, 1928), 31.

2. John J. McLaurin, *Sketches in Crude Oil* (Harrisburg, 1896), p. 341.

3. *Fort Worth Star-Telegram,* June 6, 1919.

4. Unidentified newspaper clipping in Tex Thornton scrapbook.

5. D. D. Kling, "Oil-well Shooter a Picturesque Character," *Compressed Air Magazine,* XXXVII (November, 1932), 3969.

6. Walter H. Jeffery, "Well Shooter the 'Ace' of Oil Field Workers," *National Petroleum News,* XXI (October, 1929), 226.

7. Boyce House, *Roaring Ranger* (San Antonio, 1951), p. 21.

8. Unidentified newspaper clipping, Tex Thornton scrapbook.

9. Walter Cline, tape-recorded interview, August 13, 1952.

10. *Fort Worth Star-Telegram,* June 16, 1922.

11. E. P. Matteson, tape-recorded interview, June 19, 1953.

12. Boyce House, *Were You in Ranger?* (Dallas, 1935), p. 111.

13. T. P. Byron, "Jim Hanks — Oil Shooter," *Outing,* LII (September, 1908), 685.

14. Unidentified newspaper clipping, Tex Thornton scrapbook.

15. Tex Thornton, interview, April, 1947.

16. A. H. Clough, "Dynamite Ends Gas Fire," *Petroleum Age,* VII (February, 1920), 42 ff.

17. Frank Hamilton, tape-recorded interview, July 29, 1952.

18. W. M. Hudson, Sr., interview.

19. Tex Thornton, interview.

20. The present records of the Johns-Manville Company do not show when or for whom the first asbestos suit was made.

21. Ellwood J. Munger, "Wear Asbestos: Blast Fire," *Petroleum Age,* VIII (December 15, 1921), 50.

22. For a detailed account of the putting out of this fire see L. C. E. Bignell, "Rumanian Gas Fire Finally Conquered," *Oil and Gas Journal,* XXXI (March, 1932), 16 ff. For a general article on Kinley see Stanley Frank, "He Fights the Biggest Fires," *Saturday Evening Post,* CCXXXI (May 2, 1959), 36 ff.

23. These events are covered fully by clippings to be seen in the Tex Thornton scrapbook.

24. *American Magazine*, CV (March, 1928), 26-27.

25. McLaurin, *op. cit.*, p. 352.

CHAPTER 11

1. O. G. Lawson, tape-recorded interview, July 23, 1952.

2. E. I. Thompson, tape-recorded interview, September 3, 1952.

3. Oil City (Pennsylvania) *Semi-Weekly Derrick*, March 15, 1904.

4. Walter Cline, tape-recorded interview, August 13, 1952.

5. Lawson, interview; C. C. McClelland, tape-recorded interview, July 29, 1952.

6. W. H. (Bill) Bryant, tape-recorded interview, July 29, 1952.

7. McClelland, interview.

8. J. T. (Cotton) Young, tape-recorded interview, August 15, 1952.

9. *Oil Weekly*, XXVIII, 8 (February 17, 1923), p. 96.

10. Lawson, interview.

11. Young, interview.

12. H. P. Nichols, tape-recorded interview, October 10, 1952.

13. A. B. Patterson, tape-recorded interview, August 4, 1953.

14. Cline, interview.

15. Young, interview.

16. W. J. Rhodes, interview, January 3, 1942.

17. Allen Hammill, tape-recorded interview, September 2, 1952.

18. Lawson, interview.

19. Hardeman Roberts, tape-recorded interview, April 26, 1956.

20. E. P. Matteson, tape-recorded interview, June 19, 1953.

21. McClelland, interview.

22. Young, interview.

23. Matteson, interview.

24. Lawson, interview.

25. William Edward Cotton, tape-recorded interview, May 23, 1956.

26. Lawson, interview.

CHAPTER 12

1. A. E. Ungren, interview, January 2, 1942.

2. Elizabeth K. Martin, 'The Development of the Oil and Gas Field at Premont, Texas" (unpublished MS).

3. George Fuermann, "Post Card," *Houston Post*, January 13, 1957.

4. Ed Prather, tape-recorded interview, April 14, 1954.

5. Joseph Weaver, tape-recorded interview, August 5, 1952.

6. Weaver, interview.

7. Burt E. Hull, tape-recorded interview, August 24, 1953.

8. Information supplied by the Texas Railroad Commission.

9. Bill Ingram, tape-recorded interview, July 29, 1952.

10. Information supplied by the Texas Railroad Commission.

11. G. W. Taylor, interview, July 29, 1956.

12. Quoted in Boyce House, *Were You in Ranger?* (Dallas, 1935), p. 15.

13. *New York Times,* April 28, 1922, vii, 4:2.

14. Professor Reid taught me chemistry.

15. Harry Harter, *East Texas Oil Parade* (San Antonio, 1934), p. 83.

16. John Rust, tape-recorded interview, September 12, 1952.

17. House, *op. cit.,* p. 78.

18. This story was current in Abilene, Texas, in the 1920's.

19. This story was told by a man named Stovall to a group gathered in the office of the *Herald and Press* in Palestine, Texas, July 29, 1947.

20. Quoted in House, *op. cit.,* p. 15.

21. *Ibid.*

22. Bob Duncan, *The Dickey Bird Was Singing* (New York, 1952), p. 268.

23. George Sessions Perry, *Hackberry Cavalier* (New York, 1944), p. 42.

24. Mrs. M. H. Hagarman, interview, April 6, 1942.

25. John Rust, tape-recorded interview, September 12, 1952.

26. Joseph Weaver, tape-recorded interview, August 5, 1952.

27. The story of Coal-Oil Johnny has been told many times. It is readily available in Herbert Asbury, *The Golden Flood* (New York, 1942), pp. 114-37.

28. I had this story from Richard E. Main, who heard it told by a drunk printer who had worked in Odessa, Texas.

29. House, *op. cit.,* pp. 140-41.

30. From social welfare records, Anderson County, Texas.

31. Boyce House, *Oil Boom* (Caldwell, Idaho, 1941), pp. 63-64.

32. This story was told by a Mr. Childs to a group gathered in the National Bank at Overton, Texas.

33. E. L. Lantron, tape-recorded interview, August 8, 1952.

34. L. T. Pitts, interview, April 4, 1942.

35. House, *Oil Boom,* p. 125.

36. The man from whom advice was sought was my neighbor, E. W. Everton.

37. Quoted by Elmer T. Peterson in "The Miracle of Oil," *Independent,* CXII (June 21, 1924), 340-52.

38. Richard F. Main to M. C. B., July 23, 1947.

39. "They Share the Oil but not the Toil," *Lamp,* XLII (Summer, 1960), 18-19.

40. Quoted by Peterson, *op. cit.*

CHAPTER 13

1. Carl Mirus, tape-recorded interview, April 4, 1954.

2. Hardeman Roberts, tape-recorded interview, April 26, 1956.

3. James W. Riggs, tape-recorded interview, June 19, 1952.

4. Early C. Deane, tape-recorded interview, July 7, 1953.

5. Theodore F. (Max) Schlicher, tape-recorded interview, July 1, 1953.

6. Roberts, interview.

7. Plummer M. Barfield, tape-recorded interview, August 1, 1953.

CHAPTER 14

1. *Follow de Drinkin' Gou'd,* edited by J. Frank Dobie, "Publications of the Texas Folklore Society," VII (Austin, 1928), 45-54; 55-61.

2. *Ibid.,* p. 48.

3. Theodore F. Schlicher, tape-recorded interview, July 1, 1953.

4. Claud Deer, tape-recorded interview, July 7, 1952.

5. Fred Jennings, tape-recorded interview, July 19, 1952.

6. Tape-recorded interview, September 3, 1953.

7. Tape-recorded interview, September 3, 1953.

8. Frank Shay, *Here's Audacity* (New York, 1930), pp. 63-75.

9. Carl Carmer, *The Hurricano's Children* (New York, 1937), pp. 65 ff.

10. Olive Beaupre Miller, *Heroes, Outlaws and Funny Fellows of American Popular Tales* (New York, 1942), pp. 248 ff.

11. J. H. Plenn, *Saddle in the Sky* (New York, 1940), pp. 216-20.

12. *Ibid.,* p. 216.

13. Ella Lane Carl, *The Letters of a Texas Oil Driller's Wife* (New York, 1959), pp. 153-57.

14. Plenn, *op. cit.,* pp. 153-57.

15. Carl, *op. cit.,* p. 155.

16. Bob Duncan, *The Dickey Bird Was Singing* (New York, 1952), pp. 131-56.

17. The biographical facts about Gib Morgan and the tales following are from my *Gib Morgan, Minstrel of the Oil Fields* (Austin, 1945).

CHAPTER 15

1. *Austin American-Statesman,* August 19, 1956.

2. Joseph Weaver, tape-recorded interview.

3. *Ibid.*

4. W. M. Hudson, Sr., tape-recorded interview.

5. Carl Angstadt, tape-recorded interview.

6. H. A. Rathke, tape-recorded interview, September 3, 1953.

7. Frank Bryan to J. Frank Dobie, May 23, 1940.

8. H. P. Nichols, tape-recorded interview.

Index

Accident, lucky, 62

Accidents, workers' attitude toward, 118-19

Adams, H. H.: exposes doodlebug, 40; report on "oil jumper," 21; report on "oil tromper," 21

Agricultural production, effects of oil on, 150

Alexander, Ford, uses explosives to put out fire, 112

Allen, Lew, on geology, 79

Anecdotes, 195-204; "Boardinghouse," 196-97; "Boom," 195-96; "Bulldog," 198-99; "East Texas Episode," 201-2; "Ellen Burger," 202-3; "One Ranger," 197-98; "Rotary Club," 199; "Tales of Hoffman," 200-201; "Texas Oil Millionaire," 203-4

American Magazine, quoted on Tex Thornton, 114

"Auger men," 129

Austin Statesman, quoted on lucky accident, 69

Ax, legend of, as symbol of new status, 141-44

Baker, Jake, witches for silver, 14

Bananas, purchased by newly rich, 140

Barfield, Plummer, on Negroes singing, 164

Barker, R. E., enjoins drilling in cemetery, 7

Barnett, Jackson, made wealthy by oil, 48

Big Hole Bill, driller, 126-27

Big Lake, discovery of oil near, 56, 67

Bird, Peck, wake of, 127

Blau, L. W., on doodlebugs, 36

"Bluebelly Hill," 128

Boardinghouse, anecdotes of, 196-97

"Bollweevil Song," 156

Boom, anecdotes of, 195-96

Boynton, A. P., drills in East Texas, 88

Bradford, Daisy, location of well on farm of, 60, 89

Breakdown, lucky, 63

Brooks, John Lee, on Paul Bunyan, 165

Bryan, Ruth (Madame Virginia), fortune teller, 26

Bryant, Billy: disagreement with boss, 123; on geologists, 80

"Bulldog," anecdote of, 198-99

Bunyan, Paul, as oil field hero, 165-69, 191

Byron, T. P., on oil shooter, 111

Cable tools, operation of, 128-29

Carl, Mrs. Ella Lane, on Kemp Morgan and Paul Bunyan, 172

Carmer, Carl, on Kemp Morgan, 170-71

Carruth, promotional literature of, 102

Caul, as evidence of clairvoyance, 17

Cemeteries, drilling in, 5. See *also* Graveyard

"Chain breaker," 129

Champion, Frank, 71

Chance, role of in oil finding, 58

Cheney, M. G.: experience with doodlebug, 39; on folk attitude toward geologist, 78

Clairvoyance: See Seers and X-ray-eyed

Clemmons, Spencer, reputed wealth of, 148-49